# Digital Transformation and Its Role in Progressing the Relationship Between States and Their Citizens

Sam B. Edwards III
*Quinnipiac University, USA*

Diogo Santos
*Pitágoras College, Brazil & Estácio de Sa University, Brazil*

A volume in the Advances in Electronic Government, Digital Divide, and Regional Development (AEGDDRD) Book Series

IGI Global
DISSEMINATOR OF KNOWLEDGE

Published in the United States of America by
IGI Global
Information Science Reference (an imprint of IGI Global)
701 E. Chocolate Avenue
Hershey PA, USA 17033
Tel: 717-533-8845
Fax: 717-533-8661
E-mail: cust@igi-global.com
Web site: http://www.igi-global.com



Library of Congress Cataloging-in-Publication Data

Names: Edwards, Sam B., III, 1969- editor. | Santos, Diogo, 1978- editor.
Title: Digital transformation and its role in progressing the relationship between states and their citizens / Sam B. Edwards, III and Diogo Santos, editors.
Description: Hershey PA : Information Science Reference, [2020] | Includes bibliographical references and index. | Summary: "This book examines the relationship between governments and their constituents, and how this relationship is impacted by emerging technologies"--Provided by publisher Provided by publisher.
Identifiers: LCCN 2019048577 (print) | LCCN 2019048578 (ebook) | ISBN 9781799831525 (hardcover) | ISBN 9781799831532 (paperback) | ISBN 9781799831549 (ebook)
Subjects: LCSH: Political participation--Technological innovations. | Digital communications--Political aspects. | Information technology--Political aspects. | Technology and state.
Classification: LCC JF799.5 .D55 2020 (print) | LCC JF799.5 (ebook) | DDC 323/.04202858--dc23
LC record available at https://lccn.loc.gov/2019048577
LC ebook record available at https://lccn.loc.gov/2019048578

This book is published in the IGI Global book series Advances in Electronic Government, Digital Divide, and Regional Development (AEGDDRD) (ISSN: 2326-9103; eISSN: 2326-9111)

British Cataloguing in Publication Data
A Cataloguing in Publication record for this book is available from the British Library.



For electronic access to this publication, please contact: eresources@igi-global.com.

# Advances in Electronic Government, Digital Divide, and Regional Development (AEGDDRD) Book Series

ISSN:2326-9103
EISSN:2326-9111

Editor-in-Chief: Zaigham Mahmood, University of Derby, UK & North West University, South Africa

## MISSION

The successful use of digital technologies (including social media and mobile technologies) to provide public services and foster economic development has become an objective for governments around the world. The development towards electronic government (or e-government) not only affects the efficiency and effectiveness of public services, but also has the potential to transform the nature of government interactions with its citizens. Current research and practice on the adoption of electronic/digital government and the implementation in organizations around the world aims to emphasize the extensiveness of this growing field.

**The Advances in Electronic Government, Digital Divide & Regional Development (AEGDDRD)** book series aims to publish authored, edited and case books encompassing the current and innovative research and practice discussing all aspects of electronic government development, implementation and adoption as well the effective use of the emerging technologies (including social media and mobile technologies) for a more effective electronic governance (or e-governance).

## COVERAGE

- Urban Development, Urban Economy
- Current Research and Emerging Trends in E-Government Development
- Issues and Challenges in E-Government Adoption
- ICT within Government and Public Sectors
- Frameworks and Methodologies for E-Government Development
- Case Studies and Practical Approaches to E-Government and E-Governance
- Social Media, Web 2.0, and Mobile Technologies in E-Government
- Adoption of Innovation with Respect to E-Government
- Online Government, E-Government, M-Government
- Electronic Government, Digital Democracy, Digital Government

IGI Global is currently accepting manuscripts for publication within this series. To submit a proposal for a volume in this series, please contact our Acquisition Editors at Acquisitions@igi-global.com or visit: http://www.igi-global.com/publish/.

The Advances in Electronic Government, Digital Divide, and Regional Development (AEGDDRD) Book Series (ISSN 2326-9103) is published by IGI Global, 701 E. Chocolate Avenue, Hershey, PA 17033-1240, USA, www.igi-global.com. This series is composed of titles available for purchase individually; each title is edited to be contextually exclusive from any other title within the series. For pricing and ordering information please visit http://www.igi-global.com/book-series/advances-electronic-government-digital-divide/37153. Postmaster: Send all address changes to above address. 

## Titles in this Series

*For a list of additional titles in this series, please visit:*
*https://www.igi-global.com/book-series/advances-electronic-government-digital-divide/37153*

***Cases on Electronic Record Management in the ESARBICA Region***
Segomotso Masegonyana Keakopa (University of Botswana, Botswana) and Tshepho Lydia Mosweu (University of Botswana, otswana)
Information Science Reference • © 2020 • 300pp • H/C (ISBN: 9781799825272) • US $195.00

***Employing Recent Technologies for Improved Digital Governance***
Vasaki Ponnusamy (Universiti Tunku Abdul Rahman, Malaysia) Khalid Rafique (Azad Jammu and Kashmir Information Technology Board, Pakistan) and Noor Zaman (Taylor's University, Malaysia)
Information Science Reference • © 2020 • 383pp • H/C (ISBN: 9781799818519) • US $195.00

***Political Decision-Making and Security Intelligence Recent Techniques and Technological Developments***
Luisa Dall'Acqua (University of Bologna, Italy & LS TCO, Italy) and Irene M. Gironacci (University of Swinburne, Australia)
Information Science Reference • © 2020 • 170pp • H/C (ISBN: 9781799815624) • US $195.00

***Tools, Strategies, and Practices for Modern and Accountable Public Sector Management***
Graça Azevedo (University of Aveiro, Portugal) Jonas Oliveira (Instituto Universitário de Lisboa (ISCTE-IUL), Portugal) Rui Pedro Marques (University of Aveiro, Portugal) and Augusta Ferreira (University of Aveiro, Portugal)
Information Science Reference • © 2020 • 372pp • H/C (ISBN: 9781799813859) • US $215.00

***Examining the Roles of IT and Social Media in Democratic Development and Social Change***
Vikas Kumar (Chaudhary Bansi Lal University, India) and Geetika Malhotra (Chaudhary Bansi Lal University, India)
Information Science Reference • © 2020 • 384pp • H/C (ISBN: 9781799817918) • US $195.00

*For an entire list of titles in this series, please visit:*
*https://www.igi-global.com/book-series/advances-electronic-government-digital-divide/37153*

IGI Global
DISSEMINATOR OF KNOWLEDGE

701 East Chocolate Avenue, Hershey, PA 17033, USA
Tel: 717-533-8845 x100 • Fax: 717-533-8661
E-Mail: cust@igi-global.com • www.igi-global.com

# Editorial Advisory Board

# Table of Contents

**Foreword** .......... xiii

**Preface** .......... xv

**Acknowledgment** .......... xxiv

**Section 1**
**Democratic Theory**

**Chapter 1**
The Governed and Their Governments: A Working Democratic Paradigm for the 21st Century .......... 1
*Diogo Santos, Pitágoras College, Brazil & Estácio de Sa University, Brazil*
*Mylla Maria Maria Sousa Sampaio, Universidade Federal do Maranhão, Brazil*

**Chapter 2**
John Stuart Mill's On Liberty in the Digital Communications Age: Transparency, Participation, and the Challenges of the 21st Century .......... 26
*Diogo Santos, Pitágoras College, Brazil & Estácio de Sa University, Brazil*
*Mylla Maria Maria Sousa Sampaio, Universidade Federal do Maranhão, Brazil*

**Section 2**
**Case Studies: The Americas**

**Chapter 3**
The Mexican Spring: Two Experiences on State-Society Relations in Mexico .......... 49
*Atzimba Baltazar Macias, Independent Researcher, Mexico*

**Chapter 4**
The First Amendment in the Digital Age: What Is Speech and How Should It Be Protected? ........ 68
*Sam B. Edwards, Quinnipiac University, USA*

**Chapter 5**
Online Resolution and Citizen Empowerment: Tax Appeals and Court Resolutions in North America ........ 94
*Colin Rule, Santa Clara University, USA*
*Mark James Wilson, Tyler Technologies, USA*

**Chapter 6**
In the [Source] C\ode of the Conquerors: On the Need for Culturally-Minded Tribal Law Research Databases ........ 124
*Nicholas Ravotti, Independent Researcher, USA*

**Section 3**
**Case Studies: Asia**

**Chapter 7**
Chinese Citizenry Social Media Pressures and Public Official Responses: The Double-Edged Sword of Social Media in China ........ 139
*James Masterson, Morehead State University, USA*

**Chapter 8**
Electoral Reform and the Electronic Voting System: Case of Mongolia ........ 182
*Dashbalbar Gangabaatar, Nagoya University, Mongolia*

**Compilation of References** ........ 205

**About the Contributors** ........ 227

**Index** ........ 230

# Detailed Table of Contents

**Foreword** .......... xiii

**Preface** .......... xv

**Acknowledgment** .......... xxiv

**Section 1**
**Democratic Theory**

**Chapter 1**
The Governed and Their Governments: A Working Democratic Paradigm for the 21st Century .......... 1
*Diogo Santos, Pitágoras College, Brazil & Estácio de Sa University, Brazil*
*Mylla Maria Maria Sousa Sampaio, Universidade Federal do Maranhão, Brazil*

The relationship between citizens and the state is omnipresent: even the most totalitarian regimes present some interaction between society and ruler. Digital communications have added new instruments through which all, especially the least represented, can make their voices heard. This chapter provides theoretical bases for analyzing how citizens and government interact in the digital age. Dictatorships rely on performance and resort to violence to obtain support. Democratic leaders rely on principle, performance, or elections. Legitimacy by policy efficiency can support despotic governments. Digital communications arise as new channels for governments to promote their views to the people, convincing them of their policies' value; monitor the population's activities in support or against the government; and block information contrary to governmental interests. The people too can use digital interfaces to obtain information, share approval or discontent, and organize protests. Digital communication tools deeply impact the traditional democratic mechanisms to provide regime legitimacy.

**Chapter 2**
John Stuart Mill's On Liberty in the Digital Communications Age: Transparency, Participation, and the Challenges of the 21st Century ................26
*Diogo Santos, Pitágoras College, Brazil & Estácio de Sa University, Brazil*
*Mylla Maria Maria Sousa Sampaio, Universidade Federal do Maranhão, Brazil*

On Liberty by J. S. Mill is on democracy, limitations to political power, as well as the shaping of modern democratic systems. This chapter aims to review a classic of political science in light of the current developments in the digital age and thus extract new light on the impact the new digital communication technologies have on the relationship between the state and the citizens around the world. The widespread use of social media and digital communications will have and is having deep and unavoidable impacts on the relationship between the state and citizens. However, such impacts raise both hopes of a better future for democracies and dictatorships as well as concerns of privacy, freedom of speech and thought, consistency of public policy, quality of governmental services, and even the legitimacy and lifetime of regimes. The theoretical tools of classical political science may, however, aid us in better understanding such processes and steering change for the best.

**Section 2**
**Case Studies: The Americas**

**Chapter 3**
The Mexican Spring: Two Experiences on State-Society Relations in Mexico ....................................................................................................49
*Atzimba Baltazar Macias, Independent Researcher, Mexico*

The chapter aims at understanding a recent phenomenon in Mexican politics: the use of internet and social media as a new and powerful resource for mobilization and social participation in the policy process. Based on a review of two relatively recent movements in Mexico (#YoSoy132 and The Wirikuta Defense Front), the chapter argues that although the internet is still restricted to the middle and upper classes, the use of social media and its impact transcends class boundaries, draws public attention, creates a valuable social capital for mobilization, and influences the decision-making process. The chapter does not intend to provide evidence to the theoretical discussion on why and how social media enhances political participation and mobilization; rather, it reflects on the features shared by these two movements in order to draw some lines for further research. It finds that, if used appropriately, social media is actually an effective tool to facilitate mobilization and modify the public agenda.

**Chapter 4**
The First Amendment in the Digital Age: What Is Speech and How Should It Be Protected? ....................................................................................................68
*Sam B. Edwards, Quinnipiac University, USA*

The United States is facing challenges in applying First Amendment principles from the 18th century to modern communications. This chapter examines cases where the government has intruded upon First Amendment rights. The first section examines when the government attempts to prevent protests by cutting internet access. This amounts to a digital gag and ear plugs for the protesters. Aside from cutting access to a single area, some governments have access to total internet "kill switches." This allows unsurpassed censorship of speech. Now there are technical means installed in most phones that could allow governments to disconnect internet access at the individual phone. In the area of social media, the courts are struggling to identify what constitutes speech. For example, when is a "like" or a "wink" speech? Equally important, when can a political figure censor speech through blocking users on social media? These cases represent warning signs that the United States, just like other countries, is struggling to adapt eighteenth century legal principles to modern communication.

**Chapter 5**
Online Resolution and Citizen Empowerment: Tax Appeals and Court Resolutions in North America ......................................................................94
*Colin Rule, Santa Clara University, USA*
*Mark James Wilson, Tyler Technologies, USA*

In the private sector, the growth in interactive, online technology use has already disrupted many private industries, from medicine to finance to entertainment. Interactive, online technology has empowered consumers, giving them more choices and better information, which has in turn transformed global business. Now it is clear that government services are in the process of a similar transformation. Government agencies face unique challenges in implementing interactive, online technology, and understanding best practices can be a challenge. In this chapter, the authors describe their first-hand experience helping government agencies build advanced online dispute resolution systems. They focus in particular on two case studies: transitioning property tax appeals from a paper-based process to an interactive online process and introducing interactive online technology into the courts. Through this examination, they (1) highlight the unique challenges we encountered and (2) make recommendations for government agency decision makers from the lessons we learned.

**Chapter 6**
In the [Source] C\ode of the Conquerors: On the Need for Culturally-Minded Tribal Law Research Databases ........................................................ 124
*Nicholas Ravotti, Independent Researcher, USA*

The practice of law requires not only an understanding of the law itself (i.e., what the law "says"), but also the ability to conduct proper legal research to formulate cogent legal arguments in support of one's case. For attorneys practicing before state and federal courts in the United States, this is accomplished through the use of legal research databases that catalog and archive nearly all state and federal trial court and appellate court opinions. For attorneys who practice before the 573 federally-recognized Indian tribal courts, this task is far more complex. This chapter discusses the need for a culturally-appropriate legal research database to bridge the digital divide in tribal courts.

**Section 3**
**Case Studies: Asia**

**Chapter 7**
Chinese Citizenry Social Media Pressures and Public Official Responses: The Double-Edged Sword of Social Media in China................................................ 139
*James Masterson, Morehead State University, USA*

Widespread use of social media in China is a double edged sword: social media offers opportunities for the government to connect with society, gauge the opinion of citizens in the public domain, and allow citizens to voice their anger when necessary by blowing off steam online rather than in the streets. However, social media also allows citizens to access information outside of China much more rapidly and efficiently and to link up and communicate with other citizens much more quickly. Social media allows users to share texts, photos, and files, making it much more difficult for the government to control information and to thwart organizing for political purposes. In some instances, the use of social media has forced the Chinese government to take actions that it otherwise would not have done or to reverse actions or policies already set in place. The goal of this chapter is to illustrate the double-edged sword that social media poses to government officials in China, particularly high-level party officials in Beijing.

**Chapter 8**
Electoral Reform and the Electronic Voting System: Case of Mongolia........... 182
*Dashbalbar Gangabaatar, Nagoya University, Mongolia*

Mongolia introduced a new electronic voting system for the first time for the 2012 parliamentary election. E-voting empowers citizens by making voting simpler and providing better opportunities for certain groups of citizens to participate in the election process. The electoral reform was one of the major steps the parliament carried out in order to restore public trust lost in the violent protests against the 2008 parliamentary election results. A free, transparent, and fair electoral system was important to correct the fraud in the old election system. This chapter examines the effectiveness of the mixed system of election, the electronic voting system, the constitutionality of the electoral systems, and other changes to the electoral system in Mongolia.

**Compilation of References** ............................................................................. 205

**About the Contributors** ................................................................................. 227

**Index** ............................................................................................................. 230

# Foreword

This year we celebrate the 31st anniversary of the creation of the Internet. On March 12, 1989, computer scientist Tim Berners-Lee proposed a digital information-sharing system that used linked information, nodes and hypermedia to form what he called the "World Wide Web" and now it is most commonly known simply as the "web" or the "Internet". Berners-Lee was working at a physics lab (CERN) and thought that this project would allow physicists and other scientists a platform to share information and data however, he had to undertake this project on his own time as his boss thought it to be "too vague and did not see its merits." If someone has ever been totally wrong, this is one great example.

At that time (1989) I was completing my PhD work at the University of Colorado at Boulder and remember seeing this tool that was mainly used to send electronic mail. I set up two accounts through the CU Computing Center; one for me and one for my advisor, so we could communicate on the progress of my dissertation. It worked great and it was almost instantaneous. In 1993 I became an Associate Professor at Georgia Tech and one of the first things I did was to set up my first webpage using my brand new Unix-based systems, consisting of information about my colleagues and me in the Geosystems Engineering group so we could attract the best possible graduate students to our program. It proved to be a great tool for public relations and to advertise our programs to students from around the world that had access to the Internet. This was the first time that I realized the potential power of the World Wide Web.

Since that time we have witnessed many ways in which the Web has changed the world; ways that not even its creator could have ever imagined. The Internet has been at the center of economic growth of small businesses, large corporations and countries alike, it led to great innovations in education and scientific research as well as banking and commerce, it lowered many geographic and social barriers, it changed the way and speed at which we receive news, it created the new powerful term "social media" that has led to many social and political changes empowering the citizens of the world in ways we could not even imagine only a few years ago. The Internet is intricately tied to our daily lives and today's youth could never imagine not having access to the Internet, the way us baby-boomers thought of radio and television.

However, the power of the Internet can also be abused. Recent reports of spying by government agencies (i.e. NSA, DHS, etc.) on other countries and its leaders and even on its own citizens, has raised huge concerns about the need for regulatory measures and legislation of the Internet. Hacking, cyber-terrorism and even cyber-bullying is leading to a demise of our society and its commonly held values. Dictatorships have shown to abuse the power of the Internet, often using it as a propaganda machine and to promote their self-interests. We have also witnessed of instances when dictatorships have used the Internet to monitor social unrests and punish perpetrators and "dissident activists". However, social media monitoring may also allow government entities to better understand how their people feel and perhaps be able to take corrective actions. On the other hand, democracies have also used the Internet to collect "big" data to monitor the private lives of its people and even its legislators. The big question that will arise is: will these negative aspects constrict or deter from the great contributions from the Internet? Answers to these and many other questions will come with time. In any case, the Internet is here to stay and the changes in the relationship between states and citizens will continue. We must thus try our best to understand such changes to better prepare for what is coming and also be better equipped to steer change to the directions we need/prefer, whenever that is possible. It is difficult to predict where the power of the Internet will lead us. By all means, we hope that the outcome will be a positive one.

This book deals with many of these subjects but focuses on how the Internet has led to socio-political changes in countries around the world. It details how the Internet has positively and negatively impacted governing bodies and its citizens. Each chapter presents a different story, a different country, different ways in which the Internet has impacted change around the world. The book is a compilation of articles by prominent authors and academicians from around the world, that narrate their personal accounts of how the Internet has changed the status quo and "standard" way of life in a social-political setting that makes use of instant information to disseminate to the masses. The book also details the impacts of social-media and how social-media is being used to influence decision-making and how a few people may be able to influence millions of people by blogging, tweeting, etc. events, rallies, demonstrations, protests, congregations of people interested in the same idea or a certain movement.

*Emir José Macari*
*University of New Orleans, USA*

# Preface

The Internet has brought about unprecedented access to information. Knowledge has taken on an entirely different meaning when so much information is readily available. Of course, access to this wealth of information is by no means universally available. The digital divide unfortunately separates many different groups of peoples and cultures. Despite the uneven nature of access to information, it has fundamentally changed how people and governments relate. As people are more freely able to access information they often stand up to authoritarian governments. In fact, topics in this book are dangerous to authoritarian governments. Potential authors to this book had to withdraw from this project because of fear of reprisal from their governments. Digital communications threaten governments that seek to prevent citizens from getting information, sharing information, and organizing.

The Internet has made it possible to access and share information on levels unimaginable in the past. Governments that seek to maintain power by keeping their citizens ignorant and unable to communicate are fighting a losing battle. The ability to freely communicate through the Internet is fundamentally changing the way citizens and governments interact. Although this is a losing battle it does not stop governments from trying. In fact, governments have developed internet "kill switches" where they are able to remove all users from the Internet. Manufacturers of mobile phones also have created device "kill switches" where, upon request from the government, they can cut someone's connection.

The relationship between states and their citizens is in constant evolution. Digital communication through the Internet has fundamentally changed how states relate to their citizens. Gone are the days when a government could ignore its citizens and hide its activities. Even the most autocratic governments are not immune from these fundamental changes in society. Although the interactions between the state and citizens have been of concern to both theorists and practitioners of power and government since ancient times, the changes happening today with the ability to access information, organize, and voice opinions is unprecedented. Existing theories have to be reexamined and changed to address this fundamental shift in information.

Just a few years ago, it would have been inconceivable for a political uprising to be organized as easily as it was during the Arab Spring. Comparing the governments' responses to the uprisings in the Arab Spring and Tiananmen Square demonstrates how digital communications have changed the landscape. WikiLeaks and the *I Am 132* Movement in Mexico are prime examples of how governments and traditional political parties no longer have complete control. With the ease of communication comes the ability to organize groups and to access information. This ability to organize groups and access information has changed the power balance between governments and citizens. Existing theories need to be recast in light of this fundamental change.

## OBJECTIVES OF THE BOOK

The primary objective is to examine how communication technology and the Internet has and is changing the relationship between citizens and states around the world. The ability to communicate, access information, and organize has fundamentally altered this relationship. Governments, regardless of their level of development or recognition of individual rights, must reform their policies in light of the digital revolution.

The book is designed to provide a comprehensive reevaluation of the relationship between various governments and their citizens. By providing case studies from highly developed to least developed nations we will provide a basis for comparison of this shift ongoing in societies.

## TARGET AUDIENCE

This book is appropriate as a reference book for legal professionals, activists, and government employees. It links both the technical world of digital communications with political science. Since the book has case studies from a wide range of countries, it will help provide comparative information on how various countries and citizens interact in the digital age.

## HOW THE RELATIONSHIP BETWEEN CITIZENS AND STATES IS CHANGING

The idea of a constantly connected society is one that was not even fathomable prior to the Internet. Now, however, current generations are digital natives and they cannot fathom a society that is not constantly connected. As these digital natives come into power, they will continue to press governments to change to accommodate their values related to their interconnectedness. Existing laws and legal principles dating back hundreds of years will come into conflict with the values of these new digital natives.

The fundamental legal principles underlying access to information, transparency in government, the right to share information, and the right to assemble and protest are not new, but how they are being applied to the new digital landscape is new. Governments are being faced with significant challenges as citizens gain the ability to instantly communicate with and organize large numbers of people.

Digital communications and the Internet are fundamentally changing how citizens and governments interact. These changes have positive and negative ramifications for governments and society.

## Increased Transparency

On the one hand, digital communications lead to increased transparency by governments. Citizens can much more easily know exactly what a government is doing. Even if a government seeks to hide this information from the public, the fact that the information exists in a digital form means that it can be leaked as has been proven by WikiLeaks and Edwards Snowden. Hacktivists will be able to liberate information governments seek to keep hidden to an extent that was impossible without current information technology. This increased transparency improves democracy and ultimately strengthens governments.

## Better and Different Tools

The Internet has enabled a variety of tools that can enhance or harm the relationship between citizens and states. For example, interactive, online communication enables more people to access government services. It also allows the government to better know the needs and wants of citizens. On the other hand, these tools can lead to oppression of groups and views opposed to the government. Governments can now more easily identify those who seek to oppose them. Privacy is indeed dead.

## Increased Interaction Between States and Citizens

Because of digital communications, governments and citizens can much more easily interact. Projects such as designing the 9/11 memorial and involving all of the stakeholders are now possible because the various parties can easily communicate. Governments can become more responsive to the needs of citizens such as with property tax assessment disputes. This leads to increased government efficiency.

Thus, governments can know more of what citizens want and need and more citizens can play a role in decision making. This improved communication between citizens and governments also acts to strengthen governments.

## Improved Steam Valve

Just as freedom of speech prior to the Internet, free communications are important to let the public blow off steam. This venting is important so that anger has a way of being diffused rather than increasing to the point of revolution or another dangerous outcome. The improved communications mean that citizens have an even stronger voice in which to express their feelings.

## Exponentially Larger Marketplace of Ideas

Better communications mean more exchange of information. The marketplace of information exchanged by the public means that people will be exposed to more viewpoints. It is now possible to learn so much more than when citizens were limited to print newspapers and word of mouth. This exponential growth in information through improved communications exponentially improves the marketplace of ideas. This increase in ideas in society strengthens democracy.

Although there are many advantages of improved communications through the Internet, there is also a dark side to these developments. Along with great power comes great responsibility to use it wisely. Governments, threatened with such an increase in the flow of information, sometimes abuse their power.

## There Is No More Privacy

One problem with improved technology and communications is that privacy shrinks. People now find that their every word, movement, and action can be tracked. Equally, governments are being forced to *share* more information with the public, even things that they wish to keep secret. Along with increased transparency comes increased leaks of information governments want to keep secret. For example, Julian Assange and Edward Snowden face prosecution and both are seeking refuge from friendly (to them) governments.

## Improved Ability to Protest

As citizens can assess more information and discuss with fellow citizens, they can also organize protests at a speed and on a scale that were impossible just a decade ago. Flash mobs, large movements of people, and other ways of voicing dissent threaten governments. This has and will continue to lead to instability in states. Authoritarian regimes are especially vulnerable as are states with high populations. Large movements of people create major threats.

## Technical Challenges

Governments are struggling to adapt to the new digital landscape. For example, in the United States it is no secret that the Affordable Healthcare Act, also known as Obamacare, has faced significant technical challenges as the government, for the first time, has tried to use the Inter- net for a new major program. There are also problems with too much information too poorly organized. People can access more information now than ever before but how can people trust the information they obtain? Google is necessary because no government has been able to organize its information.

## Challenges to Fundamental Rights

Technical advancements have made it possible for governments to infringe upon rights in ways that were impossible prior to the digital revolution. A government can instantly cut off all access to information with a flip of a switch, essentially silencing everyone. Governments can also access information such as every phone call and text message by everyone. Just because governments possess the technical ability to do these things does not mean that they should. Courts should protect our most fundamental rights from these technical onslaughts.

## Impacts on the Judicial System

Finally, digital communications have created new challenges on the judicial system. For example, new crimes, expanded torts such as defamation, and increases in evidence all can tax judicial systems. One rogue customer service agent can post something offensive on a company's Twitter page and instantly open the company up to liability. Class action lawsuits can be born in milliseconds. How will courts address this potential flood of cases?

## Increased Digital Divide

Not all groups benefit from these advances in communication. There is a digital divide and while more people than ever are gaining access, those left behind are even more disadvantaged. Should internet access be a fundamental right? One chapter examines how the practice of Indian law in the United States is significantly impacted from a lack of digital legal research tools taken for granted in the legal profession in the United States outside of Indian law.

## CHAPTER SUMMARIES

Chapter 1, "The Governed and Their Governments: A Working Democratic Paradigm for the 21st Century," examines the concept of democracy, one that has become indispensable for understanding the World we live on. As liberal democracies seem to be in crisis around the Globe,even the most totalitarian regimes present some interaction between society and ruler. Digital communications have added new instruments through which all, especially the least represented, can make their voices heard. Democratic leaders rely on principle, performance or elections. Legitimacy by policy efficiency can support despotic governments. Digital communications arise as new channels for governments to promote their views to the people, convincing them of their policies' value; monitor the population's activities in support of or against the government; and block information contrary to governmental interests. The people too can use digital interfaces to obtain information, share approval or discontent and organize protests.

Chapter 2, "Rereading John Stuart Mill's *On Liberty* in the Digital Communications Age: Transparency, Participation, and the Challenges of the 21st Century," reviews a classic of political science in light of the current developments in the Digital Age and thus extract new light on the impact the new digital communication technologies have on the relationship between the state and the citizens around the world. The widespread use of social media and digital communications will have and is having deep and unavoidable impacts on the relationship between the state and citizens. The theoretical tools of classical political science may, however, aid us in better understanding such processes and steering change for the best.

Chapter 3, "The Mexican Spring: Two Experiences on State-Society Relations in Mexico," examines recent phenomenon in Mexican politics: the use of Internet and social media as a new and powerful resource for mobilization and social participation in the policy process. Based on a review of two relatively recent movements in Mexico (#YoSoy132 and The Wirikuta Defense Front), the chapter argues that although the Internet is still restricted to the middle and upper classes, the use of social media and its impact transcends class boundaries, draws public attention, creates a valuable social capital for mobilization and influences the decision-making process. The chapter does not intend to provide evidence to the theoretical discussion on why and how social media enhances political participation and mobilization, rather it reflects on the features shared by these two movements in order to draw some lines for further research. It finds that, if used appropriately, social media is actually an effective tool to facilitate mobilization and modify the public agenda.

Chapter 4, "The First Amendment in the Digital Age: What Is Speech and How Should It Be Protected?" examines how constitutional doctrines including freedom of speech and assembly apply to social media in the United States. In the first section, the author examines a case where a public transportation agency was the subject of political protests. In response to the protests, the agency shut down all communications including mobile phone networks and wireless internet access. This amounted to an unconstitutional infringement on the protesters' fundamental rights of freedom of speech and assembly. Technical advancements in communications infrastructure has now made it possible for countries to shut of the Internet for all users rather than just around a single transportation station. These "kill switches" allow unsurpassed censorship on access to information and the ability to share ideas. Mobile phone manufacturers also have the ability to incorporate "kill switches" into individual phones. This allows governments to censor individual users as well. Jurisprudence has yet to catch up with these technical means for limiting access to the Internet.

In a second case study, the author examines a case where government employees were fired for "liking" a Facebook page of a political candidate. The trial court dismissed their claims holding that a "like" on Facebook did not constitute speech. The author concludes that the US legal system is struggling to adapt centuries old constitutional doctrine to the changing landscape of digital communications. Finally, when can a public figure block users on their official social media accounts? May a president block followers without violating their freedom of speech rights? The courts are trying to address this question as well.

Chapter 5, "Online Resolution and Citizen Empowerment: Property Tax Appeals in North America," discusses the efforts by public entities to implement online technology to improve their services. They focus their study on two cases: 1. property tax assessment appeals and 2. interactive online technology into the courts. These tools can enhance efficiency, empower people through access to information, and allow more people access to these essential government services. First, they examine the challenges public agencies face in using technology to interact with the public and the public's shifting expectations. Second, they discuss how the cloud model applies to public entities. In the third section, the authors examine the property tax appeals process and discuss efforts to make the process more interactive and efficient. Finally, the authors draw conclusions from the case studies and make a set of recommendations for public entities looking to update their services with online tools. In the second case study on the courts,

The introduction of these interactive tools offers many benefits to the governments and users. They also present a set of challenges. The chapter evaluates these benefits and challenges and offers solutions and paths forward for governments.

Chapter 6, "In the [Source] C\ode of the Conquerors: On the Need for Culturally-Minded Tribal Law Research Databases," examines the digital divide in the law of the many Native American court systems. Because of lack of funding and private financial incentives, much of the jurisprudence in the tribal systems is not included in digital legal research databases. The chapter also offers brief primer on federal Indian law since it can be difficult to research and is not widely understood. The chapter then details the challenges of practicing before tribal courts given this digital divide. The author concludes that this system does not adequately serve the needs of those within the jurisdiction of the tribal courts and that reform of the system is necessary to provide justice.

Chapter 7, "Chinese Citizenry Social Media Pressures and Public Official Responses," asserts that social media offers opportunities for the government to connect with society, as it gauges the opinions of citizens in the public domain and allow citizens to voice their anger when necessary by blowing off steam online rather than in the streets. Yet, social media also allows citizens to access information outside of China much more rapidly and efficiently and to link up and communicate with other citizens much more quickly. Recent developments in Hong Kong's Umbrella Movement only make long-lasting social dynamics in China more visible, as citizens try to obtain broader participation and expression spaces and the Government acts to keep democratic demand at bay. Through a series of case studies, the author examines these claims and concludes that social media helps the government pursue some of its interests at times, such as increasing nationalism and support for the government, but also has resulted in the problems for the government such as exposure of corruption and has increased public outcry.

Chapter 8, "Electoral Reform and the Electronic Voting System: The Case of Mongolia," examines the first implementation of electronic voting in Mongolia. Prior to the implementation for the 2012 elections, there was much distrust of the voting system and of the political parties. There were also violent protests in 2008 due to this distrust of the voting system. In response to the protests the government implemented electoral reforms including the introduction of e-voting. The results indicate that the electronic voting system offered several benefits such as increased proportionality, higher female representation in the parliament, and efficient elections. In addition, the electronic voting system allowed citizens residing abroad to participate in elections for the first time. The most significant benefit of the e-voting system was the restoration of popular trust in the electoral system. These were successes despite only having a short time to implement this entirely new system. The chapter then analyzes the electoral reforms of 2015 and 2019 which significantly changed the voting system yet again. Unfortunately, these reforms reversed the accomplishments of the electronic voting reforms. Despite these changes, the original reforms did help establish more trust in the voting system and contributed to constitutional development in Mongolia.

## CONCLUSION

The book examines how digital communications have and are changing the relationship between citizens and states. No country is immune from the immense effects that an interconnected society are having on governments. Tools offered by being connected both benefit and harm citizens and states. The case studies highlight the difference between common law and civil law systems. The courts of common law countries are equipped to immediately apply old doctrine to the changes made possible by advances in technology. Civil law countries face larger challenges since the changes must come from the legislative branch. The courts of common law countries are becoming the laboratories where these changes and problems are first addressed. This interconnectedness is improving and strengthening democratic institutions on one hand while also giving governments the ability to suppress and monitor the actions of people on the other. Digital communications in themselves are neither good nor bad but do present unimaginable powers to citizens and governments that were inconceivable only a decade ago.

Moore's law observes that the number of transistors on integrated circuits doubles every two years. Put another way, technology is advancing at an exponential rate. This rapid pace of technological growth presents significant challenges to existing governmental institutions. Existing laws and norms have to adapt to fundamental changes in society driven by technological advances. While technology advances at an exponential rate, the political system is not so adept at change. Moreover, because of the rapid change in technology it is difficult or impossible to predict the changes ahead. How will governments react to problems and opportunities they cannot even imagine today?

*Sam B. Edwards*
*Quinnipiac University, USA*

*Diogo Santos*
*Pitágoras College, Brazil & Estácio de Sa University, Brazil*

# Acknowledgment

We would first like to thank all of the authors who worked on chapters for this book. Not all chapters were able to be included and not all authors were able to complete their projects. Even chapters that were not ultimately included still helped advance our understanding of these important topics.

The support from our colleagues and schools (Quinnipiac University and Estácio São Luís University) helped make it possible for us to complete a project of this magnitude.

We would like to thank all of the people at IGO Global who have assisted us in this long process.

Finally, none of this would have been possible without the support of our families.

*Sam B. Edwards*
*Quinnipiac University, USA*

*Diogo Santos*
*Pitágoras College, Brazil & Estácio da Sa University, Brazil*

# Section 1
# Democratic Theory

# Chapter 1
# The Governed and Their Governments:
## A Working Democratic Paradigm for the 21st Century

**Diogo Santos**
https://orcid.org/0000-0001-9392-7689
*Pitágoras College, Brazil & Estácio de Sa University, Brazil*

**Mylla Maria Maria Sousa Sampaio**
https://orcid.org/0000-0003-1613-4604
*Universidade Federal do Maranhão, Brazil*

## ABSTRACT

*The relationship between citizens and the state is omnipresent: even the most totalitarian regimes present some interaction between society and ruler. Digital communications have added new instruments through which all, especially the least represented, can make their voices heard. This chapter provides theoretical bases for analyzing how citizens and government interact in the digital age. Dictatorships rely on performance and resort to violence to obtain support. Democratic leaders rely on principle, performance, or elections. Legitimacy by policy efficiency can support despotic governments. Digital communications arise as new channels for governments to promote their views to the people, convincing them of their policies' value; monitor the population's activities in support or against the government; and block information contrary to governmental interests. The people too can use digital interfaces to obtain information, share approval or discontent, and organize protests. Digital communication tools deeply impact the traditional democratic mechanisms to provide regime legitimacy.*

DOI: 10.4018/978-1-7998-3152-5.ch001

## INTRODUCTION

The relationship between citizens and their states is omnipresent: even in the most authoritarian or totalitarian regimes, there is always some interaction between the government and its society. Indeed, in democracies such interactions are more evident, yet they occur in every polity, even the most authoritarian ones—what makes them more difficult to notice is the rejection, by the regime, to grant a voice to the governed at least one with legitimacy or acceptability. Still, different sectors of any society will try to influence governmental decisions when they significantly affect their lives both in democracies and dictatorships (Sharp, 2010:27).

The advent and widespread use of modern digital communications has added new instruments through which all sectors of society, especially the least represented, can make their voices heard, as the latter chapters in this book will demonstrate. For this reason, understanding the relationship between the state and citizens requires revising the meaning of democracy and other concepts related to it. The purpose of this chapter is to provide a theoretical foundation for analyzing how citizens and government interact in the age of social media and digital communications, in democracies and dictatorships alike.

## WHAT IS DEMOCRACY?

Democracy is the most widely and universally accepted form government. As regimes can rarely resist for long by using force alone (Rousseau, 1913:8), it is necessary that all exercise of power be justified by one means or another so that the governed minimally accept and support such rule. A discussion of what is meant by political legitimacy is provided below in this chapter. For now, it suffices to establish that in the contemporary political debate democracy is the only source of political legitimacy that enjoys virtually absolute and complete support. This stems from the voluntarist approaches of Grotius and Locke in the $17^{th}$ century, which replaced the divine right of kings with natural law as the foundation of authority's legitimacy. Since then, the consent of the ruled is deemed as the fundamental basis for legitimate authority, as well as limitations on the power of the rulers whose authority rest in the consent of the ruled. These powers can be revoked by practical institutional mechanisms to enforce such limitations and keep the powerful in check.

This is why concepts such as democracy, checks and balances, representation and legitimacy do not have only an epistemic significance, but also a fundamental impact on what, in our age, is considered to be a democratic polity, and building a political framework to reflect such an ideal model to the best possible extent. Furthermore, as stated above, even the most authoritarian of regime leaders needs, at least to a minimal extent, the consent or recognition of her authority by the ruled. If not for legitimacy at least to aid in enforcing her authority and avoiding the need to use brute force in every social interaction, which is obviously cumbersome and ultimately impossible.

Therefore, even the most authoritarian regime must obtain a minimum legitimacy, which is demonstrated by the frequent recourse to "elections" and "referenda" in which the dictator extracts from the governed their formal consent. In the age of social media, as we will see in the latter chapters of this book, democracies and dictatorships alike use the new and widespread tools of digital communications to preach their views to their populations, share information of interest to the government, and monitor the opinions and behavior of the common people. On the other hand, populations have also used such tools to obtain relevant information to inform them- selves of their decisions (when their consent is asked or can be given), share misdeeds by the government and organize resistance and/ or protests. Democratic methods beyond the traditional representative formulas are thus absolutely relevant to analyzing the interaction between citizens and the State.

One great difficulty to our study is that one of the most contended and vaguely defined terms of political science is *democracy.* As studies on democracy have advanced, they pro- vided fewer answers than new questions (Parry & Moran, 1994:10). From its original usage in classical Greece to its modern existence, the term has been attributed and used with a myriad of senses. Indeed, finding a definition of the term democracy without qualifications has proven an elusive task. Huntington (1991:6) notes that defining democracy, either from the regime's sources of authority for the government, or by the purposes served by this government may create serious and unsolvable problems of ambiguity and imprecision. Thus, it is more feasible, and precise to define democracy from a procedural perspective, e.g. from the structures and frameworks that shape the decision-making process in a polity. In addition, a procedural definition of democracy fits better to the analysis of the State-citizenship in the digital age.

A classic definition of democracy founded in terms of its source of authority is the idea of a government based on the people's sovereignty1. This definition undoubtedly lacks precision to serve our purposes in this chapter. Perhaps it is advisable to look into the concerns of the first theorists of democracy to find what the basic problems that democratic theory intended to resolve were in order to obtain a more appropriate definition of the term. Dahl (1956:6) notes that Madison's central concern in The Federalist, No. 472, was on devising *external checks* to avoid the concentration of governmental powers, and the tyranny that necessarily results from it. Indeed, Dahl further notes that this formulation's fragility resides in the definition of its component elements: *external checks* and *tyranny*. Tyranny is defined by Madison in the same Federalist paper as "The accumulation of all powers, legislative, executive, and judiciary, in the same hands, whether of one, a few, or many, and whether hereditary, selfappointed [sic], or elective". More recently, Hart Ely notes that the one of democracy's greatest challenges nowadays is to "defend minorities from the tyranny of majorities". However circular it's argument maybe, Madison's definition is useful for this re- search, by providing us with the first element of a definition of democracy:

*Democracy is the political arrangement where the functions of government, namely, executive, judiciary, and legislative are distributed in a way to avoid their concentration and unrestricted power of one or more social actors upon others.*

Still this formulation is incomplete, for this arrangement may be used by a government with well checked powers to promote anything but the common good.

Schumpeter (1950:250) reviewed the classical theory on democracy of the eighteenth century to provide us with a definition concerned with the pursue of the common good: "democracy is the institutional arrangement for arriving at political decisions which realizes the common good by making the people itself decide issues through the election of individuals who are to [...] carry out its will". In spite of the advancement, once again we stumble on a vague term: *common good*. The difficulty here lies in determining what "common good" is at any specific time and place (ibid:250). Schumpeter proceeds then to be a critic of such a definition by noting that the mere election of individuals to realize the "common good" does not resolve satisfactorily the problem of defining it. To solve this problem, he proposes another conception of democracy in which the people elect officials to both formulate what the common good is and execute strategies to attain to it, removing from the people the responsibility to determine the constitution of its own common good in the concrete cases.

Locke, however, had already sensed that this solution in itself opens an opportunity for concentrating power. In his Second Treatise, Locke (1988 §166) defines what he calls *prerogative* as being "nothing but the power of doing public good without a rule". Kleinerman (2007:216) notes that Locke was suspicious of prerogative, as good princes could behave in way to please their people to obtain prerogative, only to later change their minds and use their concentrated powers to rule despotically. This way, a set of rules must be put in place to limit the discretion of the elected officials, besides the institutional framework that creates and enforces these rules in the first place.

Here we can arrive at a second definition of democracy:

*Democracy is the political arrangement where the functions of government, namely, executive, judiciary, and legislative are distributed in a way to avoid their concentration, and unrestricted power of one or some social actors upon others; so that to promote the common good, as defined by governing rules set by the people, to limit the discretionary powers of the rulers.*

Still, for our definition to be compete, one more aspect of democracy must be assessed, for with the definition as stated above, one problem remains: Even a definition of common good set by the legislative enshrined in the constitution and with force to check the actions of the rulers needs to be taken in a more dynamic perspective. Indeed, a static definition of a *point to be reached* is historically inconsistent, as even polities that are the most advanced in building democratic systems did not stop changing and evolving after reaching a certain level of democratization (Parry & Moran, 1994: 11).

As defended by King (2003), defining the common good requires careful and informed judgment, rather than "capricious choice or unreflective deference to prevailing opinion". One way to promote and ensure this "careful and informed judgment" (a notion close to Kant's "ideal conditions for democratic legitimacy" (Lafont, 2012)) in government is to adopt constraints to authority, requiring decisions to be broadly accepted before implemented and generating effects. More recently, Fishkin (2015) suggests three fundamental conditions for democracy: *deliberation, political equality* and *nontyranny (participation)*. *Deliberation* consists of a process where citizens can form decisions and evaluations on conflicting views during collective debates. Yet, in order for such process to have quality, it must present five elements: information, substantive balance, diversity, consciousness, and egalitarian consideration. In a nutshell, the combination of those five elements mean that available information for public debate must be precise and relevant, so that the views of one group in society may be opposed and analyzed by others, as long as there are sufficient incumbents of different views and opinions and those participating in public debate are able to ponder the merits of opposing views and their own at any moment.

As for the other two conditions, *political equity* mans that each citizen's preferences and views receive the same value, e.g. have the same mathematical weight in the general count. Finally, *participation* demands that most persons within the polity who present reasonable characteristics to be allowed to participate in public decision-making and policy-shaping processes. Most frequently such individual or group participation happens though voting, yet it may also take the form of donations, popular pressure put on incumbents through letters, email and text messages, etc., street protests, and any other form of civil society manifestation. In all fairness, Fishkin notes that building and strengthening deliberation and participation ends up weakening deliberation; since in a mass democracy, even more massified by the widespread of social media, exercising political equality and participation may lead to withering rational motivation by individuals to get informed and ponder on all relevant discussion topics. As political discussions become more frequent and intense, there is a tendency for the common citizen to converse with other likeminded individuals while avoiding debating with disagreeing parties, which hinders consideration and reflection in complex issues.

Still, building a political and institutional framework to allow and promote broad consideration and acceptance of decisions has proved to be a hard task. Robert Dahl (1971) in his theory of Polyarchies, has proposed a polity where decisions are taken in inclusive (participation of the citizenship) and contestative (openness to public opposition) frameworks. The higher the degree of inclusiveness and contestation, the closer the regime is to become a Polyarchy. In fact, our age's communication tools (social media, internet, smartphones, etc.) their affordability and consequent universality open virtually endless possibilities for the exercise of contestation and the inclusion of broad tracts of the population to the political debate, even in dictatorships. O'Donnell (1998) brings a more specific study using the concept of Polyarchies, appointing three main aspects to the constitution of a polyarchic society: A *democratic aspect*, a *republican aspect* and a *liberal aspect.*

Without going into too much detail in this interesting theoretical construction, it suffices to say that the democratic element regards to the participation of broad sectors of society, so that the making of decisions and the exercise of power are not concentrated in the hands of one or few individuals or of an elite of any kind. There are those, such as Richard Posner (2012), who criticize deliberative democracy as a mere dream, as citizens are incapable of processing the ever more complex problems of post-modern, globalized, information-age governance. Although this observation may hold some truth to it, there remains an elitist and somewhat shallow analysis of the popular consideration process. Indeed Sowell challenges the widely accepted notion that intellectuals produce better judgment on complex public problems as they individually hold and manage more information than the average citizen:

*When both special knowledge and mundane knowledge are encompassed within the concept of knowledge, it is doubtful whether the most knowledgeable person on earth has even one percent of the total knowledge on earth, or even one percent of the consequential knowledge of any given society. There are many serious implications of this, which may, among other things, help explain why so many leading intellectuals have so often backed notions that proved to be disastrous.*

In continuation, Sowell points out:

*That is why free markets, judicial restraint, and reliance on decisions and traditions growing out of the experiences of the many [...] are so important to those who do not share the social vision prevailing among intellectual elites [...] systemic processes can bring into play more knowledge for decision-making purposes, though the interactions and mutual accommodations of many individuals, than any of those individual processes.*

In this sense, it is the interactions, exchanges, debates, opposition, and conflict of ideas that fuel quality decision-making by the masses, more that the individual capability of aby of society's members.

O'Donnel's polyarchic liberal aspect refers to rights that no power, especially the State, can violate. The republican aspect of polyarchies defends that the exercise of public duties is bound by the public interest and that the actor responsible for it is subject to the limits imposed by the Rule of Law.

Comparing O'Donnell's theory to the definition we had arrived at before, we can note the same fundamental elements are present in both: First, mechanisms to limit the exercise of power by the elected politicians, in order to prevent abuse and excessive concentration of power (checks and balances). Also, present are guarantees of minimal rights and freedoms to the individuals and social actors (human rights). Finally, the promotion of the maxi- mum possible prosperity and the well being to individuals and society at large (welfare) are all fundamental and indispensable parts of a democracy.3 Among those, the missing element in our definition refers to mechanisms to assure popular participation in the political process (elections and representation, Collier & Levitsky:434).

Thus, an expansion of our definition is necessary:

*Democracy is the political arrangement where the functions of government, namely, executive, judiciary, and legislative are distributed in a way to avoid their concentration and un- restricted power of one or some social actors upon others; so that to promote the common good, as defined by governing rules set by the people to limit the discretionary powers of the rulers, who are chosen in open and competitive elections.*

This definition covers the fundamental elements of a democratic regime: mechanisms of checks and balances, the pursuit of the common good, the rule of law, and popular participation and representation. For a democratic regime do function adequately, none of those elements should be allowed to prevail over the others creating imbalances that can put the very existence of the democratic regime as such at risk.

O'Donnell (1998) points out that the democratic component (or participation) taken to its excesses may be used to justify violent and atrocious infringements of human rights against minorities or individuals on the basis that they are in the best interest of the majority: a *tyranny of the majority*. This possibility was also noticed in Dahl's formulation that "if unrestrained by external checks, any given individual or group of individuals will tyrannize over others".

The protection of individual rights and liberties of liberalism, which also apply to minorities, prevents such excesses of democracy. Still, liberalism has historically tended to lead to plutocratic4 regimes. It is the participative characteristic of the democratic tradition and the push for the *common good* and the *public interest* of republicanism that prevent the excesses of liberalism. An appropriate balance of those three traditions, so that each gives its contribution to the quality of the governance and prevents the excesses of the others, is a powerful arrangement to prevent tyranny as defined by Dahl and the risks of prerogative as exposed by Locke.

Despite which particular definition of democracy one chooses in the literature, including the one we hereby propose, the fundamental elements seem to point to the same indispensable substantial core: democracy is a method that humans use to administer their disagreements in a permanent effort for reconciling conflicting, contrasting, divergent opinions and interests in a way that different groups, or parties, can alternate on influencing the decisions and actions governing society; without neglecting the opinions or rights on any minority or opposition.

## Democratization: Transition and Consolidation

Conceptual Dependence between Democracy and Democratization

The difficulties with defining common good in a dynamic perspective have a deep impact in the concept of a democracy, as noted in the previous section. The *ideal point to be reached* is not only difficult to determine as it also changes over time within one polity.

The variety of political arrangements that can be found in the two hundred or so states in the contemporary world makes it extremely difficult to devise any concept or even classification system to comprehend clearly a significant number of states under one concept of democracy or even any classification of types of democracies. In fact, even non-democratic states may be classified under various categories, such as "authoritarian", "totalitarian", "sultanistic", "and post-totalitarian" (Linz & Stepan, 1996:27). Thus, the concept of democratization, as a process or transition from authoritarian or less democratic political arrangements to more democratic ones becomes useful to under- stand polities that although not authoritarian anymore, cannot be satisfactorily classified as consolidated democracies. The concept of democratization is also useful to orient the path ahead for those polities that wish to advance in their democratizing processes by aiding in the identification of problems to be addressed and in suggesting measures to be taken for advancement.

Basil (2004) defines democratization as:

*the "process whereby people engage in constructing a state for their own benefit with social equality as its core principle, working through an elected government operating under the rule of law, supported by functioning institutions subject to a constitutional framework incorporating international norms and standards as set out in United Nations' human rights treaties and covenants."*

This lengthy definition exposes the complexity of the underlying concept of democracy and is very close to the definition proposed in the previous section of this work. In fact, democratization could be more synthetically defined as *the process whereby a polity becomes more democratic.*

The tautology in this definition reflects the elusiveness of the concepts underlying it, in other words, as stated above, the diversity of political arrangements, expectations, cultural perspectives and conceptions of what constitutes (or should constitute) a democracy forces us to acknowledge that a closed and complete definition of democracy is impractically complex and a definition of democratization is dependent on elements relevant for the scope and perspective of each study. Therefore, the rather tautological definition of democratization adopted in this study must be understood in the light of our working definition of democracy. The most relevant point for us at this stage is perhaps how the citizenship of a dictatorship can use the tools of modern communication technology to influence more and more in the non-inclusive decision-making process of their authoritarian elites, or alas, start regime change and pursue democratization.

## Approaches and Problems Involving the Concept of Democratization

The problems involving the concept of democratization extend beyond the definition of democracy, as the literature on democracy and democratization has demonstrated that even if it was possible to determine a clearly cut and definite concept for democracy, it would be difficult to support the validity of such concept in time and space. One example of this difficulty is pointed out by Kurzman (1998:46), in which Switzerland had been consistently considered by the Polity Project to be a consolidated democracy throughout the twentieth century and has admitted women's enfranchisement only in the late 1970s. Other difficulties may also be pointed out in other elements of democracy, such as the varying levels of access to political participation and the enjoyment of human rights over time in a same polity, where levels of democratization that were perfectly acceptable in the late nineteenth century, or the early twentieth can be considered as below the minimum necessary to classify a regime as a democratic one. Huntington's influential notion of *waves of democratization* (Huntington, 1991) is rather useful to studies on democratization in providing a comprehensive theoretical framework to understand the trends of democratization and democratic regression in an impressively long time frame. However, perhaps due to the theoretical difficulties exposed above, Huntington's concept of wave of democratization is not clearly defined (Kurzman, 1998:42). As noted by Stepan (1989:viii), the conditions within which different democracies emerge and consolidate vary significantly across countries, geographical regions and time.

The amplitude of the current Huntingtonian third wave, which has allowed for democratic transitions in an unprecedented large number of polities (from southern Europe to South Africa, from Latin America to Eastern Europe and Southeast Asia) has introduced new elements and concerns relating the problem of democratization, such as how to devise and implement an institutional framework capable of accommodating the deep social fragmentation and heterogeneity in the newly democratizing polities that have emerged sometimes from authoritarianisms, in other cases from civil wars (Pildes, 2008:332). Once the initial period of democratic transition is concluded, issues of maintenance of the democratic advances and the continuations of the democratization process arise. Here, two questions may be posed. The first and relatively easy one is *what is a consolidated democracy?* The answer is rather straight- forward: *A consolidated democracy is a polity where all the relevant factors defining democracy are present, functioning and face no relevant threats of weakening or disintegration.* A more difficult question is: *Is country X a consolidated democracy, and if not, how close to consolidated is it?*

To answer this rather difficult question, a number of indices, have been devised (Hoffmann, 2005). All specific methodologies that have been developed have strong and weak points. It is not the case of finding an absolute index to measure democracy, which would be impossible, but to select the methodology that measures the relevant aspects and issues to each research objectives.

Transparency International, a global civil society organization, has developed a set of tools, or indexes, to measure the accountability and corruption levels in many countries. Its most known tool is the Corruption Perception Index (CPI), which "ranks more than 150 countries in terms of perceived levels of corruption, as determined by expert assessments and opinion surveys" (Transparency International, 2007). The main limitation of the CPI index, albeit its international credibility is that it is based on opinion polls and/ or analysts perceptions, not on empirical data. This creates the possibility of countries with higher levels of corruption to receive a better evaluation from a population, which is more tolerant to corruption.

Furthermore, the weight of opinion polls and specialist surveys is different for each country, which makes for a heterogeneous set of criteria and may compromise the precision of country rankings. The focus of the CPI index is on corruption. Such an index is useful in addressing the output aspect of legitimacy (see exposition on input and output legitimacy below) as corruption often has a direct impact on the efficiency of the State and how the citizenship perceives the usefulness of government in improving their lives. Another methodology developed to mea- sure institutional efficiency and legitimacy is the *Latinobaómetro* survey. It is a survey that collects data from interviews with representative samples of 100% of the researched countries, which are all of Latin America's, except Cuba (Latinobarómetro, 2006). The survey's scope is more appropriate, I believe, to the objective of this research, since it is more targeted to measuring the target countries' populations' perceptions on the political system and the functioning of the institutions. Examining Latinobarómetro's data may prove helpful in finding a link between the process of shaping and implementing policy in different democratizing countries and the legitimacy of the democratic system as a whole from the point of view of the population. Another strength of Latinobarómetro is that the methodology used to collect data is the same in all surveyed countries. Still, similarly to Transparency International's index, it is based on perceptions instead of more objective evidence.

Indeed, public opinion polls can be deceiving or biased, as both research data and the very public opinion can be manipulated, especially by polling institutes and the press. Yet, even the sometimes false notions broadcasted by the traditional press or social media (nowadays commonly referred to as *fake news*) are not always the creation and responsibility of journalists, and editorial board members of news channels. Such fake news can be created both consciously, by specific ill-intended individuals or groups; or unconsciously, by the collective interactions, misunderstandings, prejudices, misinformation or simply hysteria of multiple social actors. Collective perceptions on complex issues, which we examined and defended above, can also me misleading and yet enjoy considerable credibility. Abramo (2005) ponders that Transparency International's Corruption Index frequently is of interest to international investors and financiers who use its ratings to allocate investment and development aid. However, it is exactly the less developed nations, who often score poorly in the CPI the ones who receive less aid and who need it the most to escape the trap of poverty and underdevelopment. According to Larru (2009), corrupt practices do have many negative social and economic consequences, especially in developing middle and low-income countries. As investment in less developed and more corrupt countries tends to be allocated in less transparent and efficient ways, and actually creating more opportunities for corruption, the cycle poverty, corruption perception, investment and aid leading to more corruption becomes a dangerous and often inescapable trap.

One of the world's most authoritative methodologies (Hoffmann, 2005) to measure a country's level of democracy and institutional framework is Freedom House's (2007) *Freedom in The World* survey. This survey "provides an annual evaluation of the state of global freedom as experienced by individuals" classified under two broad sets: political rights and civil liberties (ibid). Still, Freedom House does not directly address the core of this research's scope, in failing to measure the quality of representation, and the institutional functioning of democratic governmental framework, albeit the consistency of the data and solidness of the methodology. A fourth authoritative index to measure the quality of democratic institutions in a country is the one used by the Polity IV Project, which presents measurements on the institutional aspects of democracy. 157 countries are rated by Polity IV considering a broad set of criteria ranging from "Democracy and Autocracy" (e.g. institutionalized democracy and regime durability) to "Characteristics of Authority" (executive constraints to defining policy, decision rules) (Marshall & Jaggers, 2005). The strong point of this databank is that it covers issues more central to this research and it is not based on opinion polls, but on more objective data.

Still, in spite of the differences among each methodology and the elusiveness of the concept of democracy, the operationalization of the different methodologies and concepts of democracy and democratization is possible. Kurzman (1998:42) has observed three main approaches to democratization, considering "levels of democracy", "net transitions", and democratization clusters or "linkages". I shall now proceed to a brief explanation of each approach in Kurzman's study:

1. **Level of Democracy:** The levels of democracy refers in this approach to the fluctuations on the number of democracies and non-democracies in the world in relation to the total number of states in each moment; this measurement may also take into account other criteria such as population. This approach adopts a quantitative, rather than qualitative approach (and in this it comes very close to Huntington's classifications), which implies classifying polities under the dual labeling of *democratic* or *non- democratic*. Although very useful to interpreting the global zeitgeist regarding democracy, this approach fails in analyzing the level of democracy within each country (e.g. the quality and functioning of democratic institutions, values and practices), in spite of its name.
2. **Net Transitions:** This approach incorporates Huntington's notion that democratization waves involve "liberalization, or partial democratization in political systems that do not become fully democratic" (Huntington, 1991:15). Although a definition to the term "fully democratic" is not provided, this approach acknowledges that different polities present different levels of democratization in a qualitative perspective. This way categories such as consolidated, non-consolidated, newly emergent and longstanding democracies can be made (Kurzman, 1998:46). Also, in this way, democratization can be viewed as a process and specific polities can be placed in a democratic gradation that ranges from non-democratic to consolidated democracy. However, the net transitions approach focuses mainly on how many countries transitioned to democracy

in relation to the number of polities that experienced democratic regressions or returned to authoritarianism (ibid:47). Still, for the purposes of research concerned with particular cases, e.g. Brazilian democratization, this approach is of limited utility.

3. **Linkages:** As noted above, the two previous approaches fail in regarding the particularities inherent to the various cases of democratizations available for study. By examining the particularities of each process of democratization, it is possible to relate different cases by the identification of similarities among cases and classifying them in groups, independently to when each polity started their transitions. In this token, Kurzman presents eight different clusters of democratizations (1998:50). This al- lows a more detailed perspective on the particularities and diversity of the many cases of democratization.

All three different approaches to the Huntingtonian perspective on democratization explained above confirm to some degree Huntington's three waves, both chronologically and regarding which countries belong to each wave. Still, approaches 2) and 3), which acknowledge a gradation in democratization, fail in providing a satisfactory response to what a full democracy is. In this respect, we return to the beginning of this discussion, and reaffirm the utility of having a working definition of democracy, useful for specific purpose and dependent on the specific objectives of a research. In that way, our working definition of democracy, as developed above, can be combined with the tautological notion of democratization as the process whereby a polity becomes more democratic to the following definition of democratization:

*Democratization is the process whereby a polity devises, implements, consolidates and improves a political arrangement where the functions of government, namely, executive, judiciary, and legislative are distributed in a way to avoid their concentration and unre- stricted power of one or some social actors upon others; so that to promote the common good, as defined by governing rules set by the people to limit the discretionary powers of the rulers, who are chosen in open and competitive elections.*

Needless to say, this concept of democratization shares the same limitations and incompleteness as the definition of democracy upon which it was built. This is due to the specific purposes of this study, which focus on the procedural and institutional aspects of democracy, for the institutional and legal structures that organized a democratic regime have a decisive importance on shaping the practice and experience of democracy in a given polity (Pildes, 2008:319). Other important aspects that have been left aside in this definition, but that are of utmost importance for this research, such as the problems of legitimacy and the exercise of different governing functions by different branches of government will be covered in the following sections.

## LEGITIMACY: INPUT AND OUTPUT

Another fundamental and rather vexing concept of political science is that of legitimacy, which has been preliminarily examined in the beginning of this chapter and to which we will dedicate more attention now. Legitimacy can be defined as the foundation upon which any state claims to exercise power or control over society (Ansell, 2001:8704) as well as instrumental to the maintenance of any regime namely democratic or democratizing or authoritarian. Foucault (1991:102) gives us one fundamental view on the phenomenon of legitimacy in his definition of *governmentality*:

*The ensemble formed by the institutions, procedures, analyses and reflections, calculations and tactics that allow the exercise of this very specific albeit complex form of power, which has its target population, as its principal form of knowledge political economy, and as its essential technical means apparatuses of security.*

In this way, Foucault acknowledges three elements for *governmentality*: a structural one (institutions), a functional one (procedures) and a cognitive one (analyses, reflections, calculations and tactics) (Brinkerhoff & Goldsmith, 2002:1). Foucault also points out the teleological aspect of governance in this definition, that is, to administer society. Furthermore, the state also uses its apparatuses of security (legitimate use of force) to implement the ideas and projects devised (Polito, 1994). In this token, the use of force by the state needs legitimacy. However, the idea of legitimate use of force was not originally formulated by Foucault, but by Max Weber (1947:78), in his defined "state" as being the "human community that (successfully) claims the *monopoly of the legitimate use of physical force* within a given territory". The concept of legitimacy is, thus, in the core of any claim to authority, made by any government, state, or human authority which wished to be obeyed (Ansell, 2001:8704). For Weber, the legitimate use of force is not only a necessary condition for the maintenance of a political regime, but it is also indispensable to the very existence of the state (Weber, 1947:78):

*Like the political institutions historically preceding it, the state is a relation of men dominating men, a relation supported by means of legitimate (e.g. considered to be legitimate) violence. If the state is to exist, the dominated must obey the authority claimed by the powers that be.*

Weber's wording between the parentheses "considered to be legitimate" touches a central point to the understanding of legitimacy: that legitimacy depends on a number of subjective criteria and is closely related to the recognition of authority by the ruled. Beetham (1993:488) argues that the rule of the powerful can be legitimized from three criteria, namely: (1) that power should be acquired and exercised according to established rules; (2) that the rules should be justifiable by reference to shared beliefs; (3) and there should be appropriate actions expressive of consent on the part of those qualified to give it. In a nutshell, democratic systems find their legitimacy on the observance of a number of principles which may variate according to each polity, but that also find their minimal configuration in the government's accountability and transparency towards their citizens, in that laws are administered equitably and fairly and that all citizens have access to the political process (Anderson & Tverdova, 2003:91).

*The principles underlying democratic political systems presume that governments are ac- countable to their citizens, that they administer laws equitably and fairly, that their actions are transparent, and that all citizens have access to the political process.*

Furthermore, Weber maintains that authoritarian regimes have limited options for obtaining legitimacy, and have to resort on performance to obtain support. Democracies, on the other hand, can usually resort on other sources of support, such as the principles be- hind the establishing of the regime itself and governmental procedure (constitutionalism) and the legitimacy brought by popular election (representation).

Observance to such principles can create legitimacy to a regime based on the regime's compliance with Beetham's shared beliefs and the consequent consent given by the ruled. However, Rogowski, cited by Ansell (2001:8706) also notes that there can be legitimacy based on more pragmatic reasons; as noted by Olson (1991:131), authoritarian rulers have a sharp interest in promoting development within their domains for this makes the ruler herself more powerful. Also, tyrants who improve the living conditions and prosperity of their peoples are likely to have their authority and support reinforced (Locke, 1988:§107 and §161). Thus, a benevolent dictator would be preferable to an inefficient democracy. However, Locke further notes that the people's focus on governmental efficiency in promoting welfare allows the ruler to feel increasingly at will to pursue his own interests, tending to act to limit the people's freedoms whenever these clash with her immediate interests (ibid:§166). In order to articulate the relationship between governmental efficiency and restraints to the exercise of power for matters of principle legitimacy may be perceived under two complementary perspectives (Kersbergen & Waarden, 2004:156; Kaufman et. al., 1999:5):

1. **Input Legitimacy:** The appropriateness of the procedures used in the policy shaping and implementation in light of the broader set of values adopted in that polity. In a state that is democratic or is in the process of implementing/ consolidating democracy, decisions must be made and implemented according to minimal procedural standards6 of accountability and transparency: the failure by governments to meet these standards tends to create a perception that such governments are arbitrary, despotic and unaccountable.
2. **Output Legitimacy:** Results from the efficacy of the policy in addressing the issue at stake; whether the policy is ad- equate to resolve the problems, deliver the desired goods/services or create the intended benefits to society. When policies are not able to rule and organize social interactions "or to pursue any kind of collective interest in an authoritative fashion" (Moore, 2001:2) they may be perceived as inadequate by the population and therefore lacking output legitimacy.

The legitimacy brought by policy efficiency can support despotic governments over time and even allow for very well-off societies. However, in the longer term, lack of controls on the actions of a ruler exposes the whole polity to the risk of abuse of power and deterioration of the people's liberty. In a democratic polity, decision-making for economic and social development must occur within the framework of accountability and transparency provided by the democratic method (see Shumpeter and Dahl's theories discussed above) (Grindle, 2007:554-555).

In more specific terms, the elements constituting each type of legitimacy can be listed as:

1. **Separation of Powers:** The division of governmental functions into three branches of government—Executive, Legislative, Judiciary.
2. **Checks and Balances:** Mutual controls by each branch of government on the exercise of power by the other branches in order to avoid the concentration of political power. Under this principle of government, separate government branches are empowered to prevent ac- tions by other branches and are induced to share power (Checks and Balances, 2008).
3. **Representation:** Governing functions are exercised by individuals chosen by the people in competitive elections.

4. **Rule of Law:** The actions of the elected rulers and the state are limited by a pre- assigned set of regulations which are defined by the people, through its representatives, or directly in order to prevent the use of political power to oppress or restrict the liberties of the individuals.
5. **Pursuit of the Common Good:** The ultimate goal of the state's actions, the common good in concrete terms is also defined by the representatives of the people and expresses the objectives and aspirations of each society. **Governmental Efficacy:** Refers to the extent to which the policies adopted by the government achieved their respective objectives; in other words, efficacy is related to the results of governmental action to promote the common good as defined in item 5) above.

The elements regarding Input Legitimacy vary greatly according to the political culture and values of each society. In democratic regimes and societies, such values are associated with the principles of constitutionalism and democratic theory ingrained in our definition of democracy, and which were discussed above and refer to items 1) to 4) in the list above. Output legitimacy, however has a more practical character, and varies less among different political cultures, namely between democracies and dictatorships and are listed in items 5) and 6) above.

Still, input and output legitimacy are not isolated and independent from each other, as the elements pertaining to each type of legitimacy (items "1" to "6" above) may im- pact legitimacy in both aspects. According to the World Bank (n.d.:7), assessing quality of governance (output legitimacy) requires attention to three areas mainly: (1) Selection, accountability and replacement of authorities; (2) Efficiency of institutions, regulations, resource management–Regulatory frame- work and Government effectiveness; and (3) Respect for institutions, laws and interactions among players in civil society, business, and politics–control of corruption and rule of law.

Of those, the two latter ones are relevant for our assessment: The constitution and shaping of the institutional framework is a fundamental aspect of input legitimacy (more evident in separation of powers, checks and balances, representation and rule of law aspects, but also in other minor ones not mentioned here) (Pildes, 2008:332), furthermore, this institutional framework's functioning has a direct impact of the output legitimacy records of a regime, and in Brazil provisional measures have been used as a way to obtain policy efficiency by the executive in a politically fragmented polity with a slow and undecided parliament, as will be discussed in chapters 3 and 4 below.

The input aspect of legitimacy of a polity is also fundamentally impacted by the institutional quality of its political frame- work: According to O'Donnell (n.d.:6-7), the governmental institutional framework of a polity provides social actors with (1) a framework for decision-making, (2) necessary patterns of representation, (3) the stabilization of expectations by agents and a longer time-horizon for social actors to plan and execute their aspirations. In addition to that, the third area mentioned by the World Bank (institutions and interactions among actors in society) may be harmed by the extensive use of unilateral executive acts to shape and implement policy. The rule of law is not thus an element only important to the input aspect of legitimacy: According to the International Monetary Fund (1997:v) ensuring the rule of law and improving the accountability of the public sector are essential elements of a legal and institutional environment in which economies can prosper (output legitimacy).

Checks and balances also have an impact in both the input and the output aspects of governance and require that the separate branches of government to be empowered to limit and control the actions of the others (Checks and Balances, 2008), while retaining a minimum degree of autonomy to enable them to perform their institutional functions. The mutual and self-limiting action exercised by each branches on the others is intended to prevent abuse and concentration of power. Therefore, verifying the extent to which policy has been shaped and implemented by decree, passing congressional checks, in diverse policy areas, is essential for addressing the quality of the rule of law, separation of powers, checks and balances and representation in Brazil, especially the extent in which the adoption of rules and policies is stable, transparent and understandable by the civil society and private sector; as these aspects have important and deep impacts in the input and output legitimacy of the Brazilian democratizing regime.

## THEORIES ON DELEGATION AND UNILATERAL ACTION

As seen in the previous section, good governmental performance may be generated by democracies or dictatorships (output legitimacy), but only a working system of checks and balances and representation (input legitimacy) can guarantee rights and liberties to citizens in the longer term. In light of this, we may note that the wise government by a benevolent dictator, who uses her power to promote the common good and the progress and well- being of society by increasing governmental efficiency and performance would harvest the support of large tracts of the citizenship (Output Legitimacy). However, the lack of inclusiveness in the decision-making process and the resulting concentration of powers in the hands of the benevolent dictator expose the population to the constant use of force to maintain the power of the leader.

As will be discussed later in this book, the current context of growing discontent with the performance of the democratic and dictatorial regimes around the world poses pressure on politicians to deliver results in macroeconomic administration, increase the general living standards of their constituencies and improve the extent and quality of public services (Veloso, 2006). Such discontent has been increasingly formulated and expressed, by the use of digital social media, which has been changing the manner in which the state and citizens interact in democracies and dictatorships worldwide. This has to do with obtaining and keeping output legitimacy, a fundamental task for politicians, especially for the chief of the executive branch.

This urge for performance validation (out- put legitimacy) by the government puts the government in a delicate position and creates incentives for using decision-making procedures with a focus on results, at the expense of the appropriateness of procedure. Indeed, the greater the head of government's scope of powers, and the lower the limitations on the use of power by an immobile parliament (wherever there is one), the greater the chance for the leader to govern alone. Also, the effort to assure output legitimacy may jeopardize the regime's input legitimacy with the use of power to the constitutional limit by the leader (Skach, 2005). Carey and Shugart (1998:2) note that the intensive use of executive decree powers is frequently perceived as evidence of inefficacy of democratic institutions and marginalization of legislatures.

As Pereira et al (2005:178) note, the transference of policy-making powers from the legislatures to the executive was a trend observable in democratic regimes throughout the twentieth century. The exacerbation of such a trend has led Przeworsky to observe that, especially in Latin American countries, "the concentration of powers in the hands of the president has reduced the political process to a succession of elections, executive decrees and sporadic outburst of protest" (Przeworsky,1991:186).

However, the mere greed for power of chiefs of executives or the apathy of legislatures do not seem to be good enough reasons to explain why policy-making powers have tended to concentrate in the hands of the chief executive in democratic regimes. The literature has provided two main explanations for increasing presidential decree powers (Pereira et al., 2005:180):

First is the "Unilateral Action Theory," which is the most traditional of those approaches to presidential policy-making by decrees. Presidents (or Prime-ministers) have the primary interest and are the most benefited by decrees. Presidential decrees are used by presidents to bypass uncooperative or adversarial legislatures, acting in isolation to make policy on their own, thus altering the *status quo* to fit the president's interests.

Unilateral action explanations do not rely simply on presidential desire to define policy alone, but its defenders see a relationship between decree usage and congressional support, in which presidents will use their decree powers more often when facing strong opposition or weak support in the legislative. However, Mayer (1999:462), an advocate of unilateral action, has found that, contrary to his hypothesis, in the United States, a divided government (i.e. when the ruling party or coalition in Congress is different to that of the president) will produce fewer decrees than situations when the president held a ruling majority in the legislative.

Second is the "Delegation Theory," which defends that, not only the president, but also legislators also benefit from presidential decrees. As Carey & Shugart (1998:296) put it, "executive decree can be attractive to legislators as a means of expediting action on policy, even without ceding control over policy to the executive". The main evidence to this is that congressmen hold considerable legislative power to overrun and reverse policy choices made by the president if they so wish (Pereira et al. 2005:181). In fact, decree powers are given to the executives by legislatures, and can be removed or overrun with relative ease, and still presidents keep holding their decree prerogatives and exercising them more or less freely.

Either delegation or unilateral action theories provide a satisfactory explanation to the causes of extensive and growing use of decrees in presidential systems observed in the past decades. A more comprehensive approach, including perhaps elements and contributions from both theories, may probably provide a more elucidating explanation. Whether decree-made policy derives from presidential greed for power, or congressional delegation, other factors have to be taken in consideration.

## CONCLUSION

The diversity of presidential regimes, with their myriad of institutional designs, authoritarian traditions (or not), and policy-making frameworks demand a comparative outlook into the main presidential regimes in the western hemisphere, with a history of executive preponderance over the other branches of government and authoritarianism. In dictator- ships, the leader or the small elite that controls society are also under pressures to promote the wellbeing of their populations or resort to an increasingly cumbersome use of violence to maintain the obedience of their governed. In this context, the tools of digital communication arise as new channels for governments to promote their views to the people, convincing them of their usefulness and value; monitor the population in general or specific sectors' activities in support or against the government and to filter or block information contrary to the interests of the government, the dictator or the ruling elite. The population also has numerous new ways to use the new digital interfaces in its relationship with the State: obtain information relevant to decision-making, share approval or discontent to other sectors of society, organize protests or support among others. All in all, for both the State and the citizens, and in both democracies and dictatorships, the traditional mechanisms of democracy to provide legitimacy to the regimes are deeply impacted by the new communication technologies, as the following chapters of this book are to demonstrate.

## REFERENCES

Abramo, C. W. (2010). Percepções pantanosas: a dificuldade de medir a corrupção. *Revista Novos Estudos, 34*(3).

Anderson, C. J., & Tverdova, Y. (2003). Corruption, Political Allegiances, and Attitudes Toward Government in Contemporary Democracies. *American Journal of Political Science, 47*(1), 91–109. doi:10.1111/1540-5907.00007

Ansell, C. K. (2001). Legitimacy: Political. In *International Encyclopedia of the Social & Behavioral Sciences* (pp. 8704–8706). New York: Elsevier. doi:10.1016/B0-08-043076-7/01167-0

Basil, W. J. (2004). Democratization: Transitions and Conflicts in Asia and the Pacific. *Asian Human Rights Commission*. Retrieved June 30, 2009 from: http://www.ahrchk.net/ pub/mainfile.php/pandp/164/

Beetham, D. (1993). In Defence of Legitimacy. *Political Studies, 41*(3), 488–491. doi:10.1111/j.1467-9248.1993.tb01651.x

Brinkerhoff, D. W., & Goldsmith, A. A. (2002). *Clientelism, Patrimonialism and Democratic Governance: An Overview and Framework for Assessment and Programming*. Retrieved August 08, 2007 from http://www.abtassoci- ates.com/reports/2002601089183_30950.pdf

Carey, J., & Shugart, M. (Eds.). (1998). *Executive Decree Authority*. Cambridge, UK: Cambridge University Press; doi:10.1017/ CBO9780511609305

Checks and Balances. (2008). In *Encyclopædia Britannica*. Retrieved July 03, 2008 from: http://www.britannica.com/Ebchecked/ topic/108283/checks-and-balances

Collier, D., & Levitsky, S. (1997). Democracy with "Adjectives". *World Politics, 49*(03), 430–451. doi:10.1353/wp.1997.0009

Dahl, R. (1956). *A Preface to Democratic Theory*. Chicago: The University of Chicago Press.

Dahl, R. (1971). *Polyarchy*. New Haven, CT: Yale University Press.

Ely, J. H. (2010). *Democracia e Desconfiança; uma Teoria do Controle Judicial de Constitucionalidade*. São Paulo: WMF Martins Fontes.

Fishkin, J. S. (2015). *Quando o Povo Fala: Democracia Deliberativa e Consulta Pública*. Curitiba, Brazil: Instituto Atuação.

Foucault, M. (1991). Governmentality. In G. Burchell, C. Gordon, & P. Miller (Eds.), *The Foucault Effect, Studies in Governmentality*. Chicago: The University of Chicago Press.

Freedom House. (2007). *Freedom in the World*–2007. Retrieved December 06, 2007 from: http://www.freedomhouse.org/template. cfm?page=351&ana_page=333&year=2007

Grindle, M. S. (2007). Good Enough Governance Revisited. *Development Policy Review*, *25*(5), 553–574. doi:10.1111/j.1467-7679.2007.00385.x

Hamilton, A. J. J., & Madison, J. (1787-88). *The Federalist Papers*. Retrieved February 20, 2014 from: http://www.gutenberg.org/ etext/18

Hoffman, A. (2005). Political Parties, Elector- al Systems and Democracy: A Cross-national Analysis. *European Journal of Political Research*, *44*(2), 231–242. doi:10.1111/j.1475-6765.2005.00225.x

Huntington, S. (1991). *The Third Wave: Democratization in the Late Twentieth Century*. Norman, OK: University of Oklahoma Press.

International Monetary Fund. (1997). *Good Governance: The IMF's Role*. Retrieved January 30, 2007 from: http://www.imf.org/ external/pubs/ft/exrp/gvern/govindex.htm

Kaufmann, D., Kraay, A., & Zoido-Lobaton, P. (1999). Aggregating Governance Indicators. *World Bank Policy Research Working Paper No. 2195*. Retrieved November 15, 2007 from: http://www-wds.worldbank.org/ex- ternal/default/WDSContentServer/IW3P/IB/1999/10/23/000094946_99101105050593/additional/115515322_20041117135531.pdf

Kersbergen Kees, V., & Waarden, V. (2004). 'Governance' as a Bridge Between Disciplines: Cross-disciplinary Inspiration Regarding Shifts in Governance and Problems of Governability, Accountability and Legitimacy. *European Journal of Political Research*, *43*(2), 143–171. doi:10.1111/j.1475-6765.2004.00149.x

King, L. A. (2003). Deliberation, Legitimacy and Multilateral Democracy. *Governance: An International Journal of Policy, Administration and Institutions*, *16*(1), 23–50. doi:10.1111/1468-0491.t01-1-00203

Kleinerman, B. A. (2007). Can the Prince Really be Tamed? Executive Prerogative, Popular Apathy, and the Constitutional Frame in Locke's Second Treatise. *The American Political Science Review*, *101*(2), 209–222. doi:10.1017/S000305540707013X

Lafont, C. (2012). Agreement and Consent in Kant and Habermas—Can Kantian Constructivism be Fruitful for Democratic Theory? *The Philosophical Forum*, *43*(3), 277–295. doi:10.1111/j.1467-9191.2012.00425.x

Latinobarómetro. (2006). *Latinobarometro Report: Online Databank—2006*. Retrieved December 03, 2007 from: http://latinobaro- metro.org/index.php?id=101

Larru, J. (2009). Corrupción y ayuda al desarrollo. Evidencias, teoría y aplicaciones para España. *Cuadernos Económicos de ICE*, *78*, 261–280. doi:10.32796/cice.2009.78.5976

Linz, J. J., & Stepan, A. (1996). *Problems of Democratic Transition and Consolidation: Southern Europe, South America, and Post-Communist Europe*. Baltimore, MD: Johns Hopkins.

Locke, J. (1988). *Two Treatises of Government*. New York: Cambridge University Press; doi:10.1017/CBO9780511810268

Marshall, M., & Jaggers, K. (2005). *Political Regime Characteristics and Transitions, 1800-2004— Polity IV Project Dataset User's Manual.* Retrieved December 07, 2007 from: http://www.cidcm.umd.edu/polity/data/pol- reg.htm

Moore, M. (2001). Political Underdevelopment: What Causes Bad Governance? *Public Management Review*, *1*(3), 385–418. doi:10.1080/14616670110050020

O'Donnell, G. (1998). Horizontal Account- ability in New Democracies. *Journal of Democracy*, *9*(3), 112–126. doi:10.1353/jod.1998.0051

O'Donnell, G. (n. d.). *Delegative Democracy?* Kellogg Institute Working Paper. Retrieved July 12, 2007 from: http://kellogg.nd.edu/ publications/workingpapers/WPS/172.pdf

Olson, M. (1991). Autocracy, Democracy and Prosperity. In R. J. Zeckhause (Ed.), *Strategy and Choice*. Cambridge, MA: MIT Press.

Parry, G., & Moran, M. (Eds.). (1994). *Democracy and Democratization*. New York: Routledge; doi:10.4324/9780203198551

Pereira, C., Power, T. J., & Rennó, L. (2005). Under What Conditions do Presidents Re- sort to Decree Power? Theory and Evidence from: The Brazilian Case. *The Journal of Politics*, *67*(1), 178–200. doi:10.1111/j.1468-2508.2005.00313.x

Pildes, R. (2008). The Legal Structure of Democracy. In *Whittington, Keith; Kelemen, Daniel & Caldeira, Gregory. The Oxford Handbook of Law and Politics*. Oxford, UK: Oxford University Press. doi:10.1093/oxfordhb/9780199208425.003.0018

Polito, M. (1994). Wit, Will and Governance in Early Modern Legal Literature. *Mosaic*, *27*(4), 15–34.

Posner, R. (2012). *A Problemática da Teoria Moral e Jurídica*. São Paulo: Editora WMF Martins Fontes.

Rousseau, J. J. (1913). *The Social Contract*. London: J.M. Dent & Sons.

Schumpeter, J. (1950). *Capitalism, Social- ism and Democracy*. New York: Harper & Brothers.

Sharp, G. (2010). *Dictatorship to Democracy—A Conceptual Framework for Liberation*. Boston: The Albert Einstein Foundation.

Skach, C. (2005). Constitutional Origins of Dictatorship and Democracy. *Constitutional Political Economy*, *16*(4), 247–368. doi:10.100710602-005-4316-x

Sowell, T. (2011). *Intellectuals and Society*. New York: Basic Books.

Transparency International. (2007). *Transparency International Corruption Perceptions Index: A Short Methodological Note*. Retrieved January 15, 2008 from: http:// chapterzone.transparency.org/content/down- load/23966/358199

Veloso, G. R. (2006). *Clientelismo: Uma Instituição Política Brasileira* (MA dissertation). Universidade Nacional de Brasília. Retrieved April 04, 2008 from: http://repositorio.bce. unb.br/handle/10482/1431

Weber, M. (1947). *From Max Weber: Essays in Sociology*. London: Gerth & Mills.

World Bank. (n. d.). *Good Governance and Its Benefits on Economic Development: An Overview of Current Trends*. Retrieved January 30, 2007 from http://www. worldbank.org/ wbi/governance/pdf/eduardpresentation_mu- nich_inwent.pdf

## KEY TERMS AND DEFINITIONS

**Checks and Balances:** Mutual controls by each branch of government on the exercise of power by the other branches in order to avoid the concentration of political power. Under this principle of government, separate government branches are empowered to prevent actions by other branches and are induced to share power (Checks and Balances, 2008).

**Common Good:** The ultimate goal of the state's actions, the common good in concrete terms is also defined by the representatives of the people and expresses the objectives and aspirations of each society.

**Democracy:** Refers to the political arrangement where the functions of government, namely, executive, judiciary, and legislative are distributed in a way to avoid their concentration and unrestricted power of one or some social actors upon others; so that to promote the common good, as defined by governing rules set by the people to limit the discretionary powers of the rulers, who are chosen in open and competitive elections.

**Legitimacy:** The source of the State's claim to power over society, and a necessary condition for the continued maintenance of any established political regime.

**Representation:** Governing functions are exercised by individuals chosen by the people in competitive elections.

**Rule of Law:** The actions of the elected rulers and the state are limited by a pre- assigned set of regulations which are defined by the people, through its representatives, or directly in order to prevent the use of political power to oppress or restrict the liberties of the individuals.

**Separation of Powers:** The division of governmental functions into three branches of government—Executive, Legislative, Judiciary.

## ENDNOTES

1. Lincoln, Abraham. (November 19, 1863). *The Gettysburg Address.* Library of Congress. Retrieved (2009/04/17): http://www.loc.gov/exhibits/gadd/ga-drft.html
2. Hamilton et al. (1787-1788). The Federalist Papers. Retrieved (2009/07/22): http://www.gutenberg.org/etext/18
3. These elements can be found, more or less frequently, in studies on democracy. Some of them: Smith (1991: 466-467), Huntington (1991:7), Linz & Stepan (1996:Part I), Parry & Moran (1994:3-8), Robinson (1998:150), Revel (1993:3).
4. Plutocracy: The rule of the wealthy, government by an economical and/or financial elite.
5. Zeigteist is a german-origined word that can be defined as the "general intellectual, moral, and cultural climate of an era". ZEITGEIST. (2010). In *Merriam- Webster Online Dictionary.* Retrieved (2010/01/03): http://www.merriam- webster.com/dictionary/ZEITGEIST)
6. For the "minimal standards" of account- ability and transparency prevailing in the Brazilian democratic culture and legal system, refer to section 4.1.2 below.

# Chapter 2
# John Stuart Mill's On Liberty in the Digital Communications Age:
## Transparency, Participation, and the Challenges of the 21st Century

**Diogo Santos**
https://orcid.org/0000-0001-9392-7689
*Pitágoras College, Brazil & Estácio de Sa University, Brazil*

**Mylla Maria Maria Sousa Sampaio**
https://orcid.org/0000-0003-1613-4604
*Universidade Federal do Maranhão, Brazil*

## ABSTRACT

*On Liberty by J. S. Mill is on democracy, limitations to political power, as well as the shaping of modern democratic systems. This chapter aims to review a classic of political science in light of the current developments in the digital age and thus extract new light on the impact the new digital communication technologies have on the relationship between the state and the citizens around the world. The widespread use of social media and digital communications will have and is having deep and unavoidable impacts on the relationship between the state and citizens. However, such impacts raise both hopes of a better future for democracies and dictatorships as well as concerns of privacy, freedom of speech and thought, consistency of public policy, quality of governmental services, and even the legitimacy and lifetime of regimes. The theoretical tools of classical political science may, however, aid us in better understanding such processes and steering change for the best.*

DOI: 10.4018/978-1-7998-3152-5.ch002

## INTRODUCTION

The world is in constant change. The relationships between the many social actors are also in permanent mutation and redefinitions. What variates over time is the nature of such changes, their causes and possible outcomes. As each phase in human history proposes different challenges, specific solutions must be devised, or unknown consequences must be faced. Our time's new digital communication technologies have impacted the relationship between the State and citizens, as the title of this book suggests, but the real questions are of a deeper nature. Shifts in the relationship between the State and the people call for a re-shaping of political science and some dogmas on the functioning of Law and the State. In order to provide answers to such questions, we may turn to the classics of political science—those timeless works that survive the vicissitudes of time and that are therefore capable of providing new answers if read with new eyes.

*On Liberty* (Mill, 1989) is about democracy, the limitations to political power and to the actions of individuals that protect the ability of the people to pursue their dreams and aspirations as well as the shaping of modern democratic systems. At the time it was writ- ten, intellectuals in the advanced democracies could ignore what was happening in other parts of the world. This is no longer the case, as the extensive and profound integration of the globalized world and the consequent shared challenges are forcing the search for common answers (to shared questions) to advanced de- mocracies, young and consolidating ones and dictatorships alike. Today, we cannot afford devising answers and explanations to just one (small) part of the so-called "civilized world".

The objective of this chapter is to review a classic of political science under in the light on the current processes and developments in the Digital Age and thus extract new light on the impact the new digital communication technologies are affecting the relationship between the State and the citizens around the world.

Some of the the questions that must be asked can be extracted from the letter by Mark Zukerberg, co-founder and CEO of *Facebook* to the United States Securities and Exchange Commission (Zukerberg, 2012):

*[…] We hope to change how people relate to their governments and social institutions.*

*We believe building tools to help people share can bring a more honest and transparent dia- logue around government that could lead to more direct empowerment of people, more ac- countability for officials and better solutions to some of the biggest problems of our time.*

*By giving people the power to share, we are starting to see people make their voices heard on a different scale from what has historically been possible. These voices will increase in number and volume. They cannot be ignored. Over time, we expect governments will become more responsive to issues and concerns raised directly by all their people rather than through intermediaries controlled by a select few.*

*Through this process, we believe that lead- ers will emerge across all countries who are pro-internet and fight for the rights of their people, including the right to share what they want and the right to access all information that people want to share with them.[...]*

**Question 1:** Is the relationship between governments and their peoples changing?

**Question 2:** What impact the new social media and digital communication technologies have in improving transparency and accountability by governments to their peoples?

**Question 3:** How are the roles of the representation system (in democracies) and the State bureaucracy (in both democracies and dictatorships) impacted by the new possibilities of direct participation, information gathering and sharing, and mobilization brought by the new social media and digital communication technologies? Mr. Zuckerberg's words suggest that his answers to such questions are positive and optimistic, as fits the justification letter by a company's CEO to its first IPO. However, answering them is no easy feat and the objective of this chapter is to use the traditional theory of political science expressed in John Stuart Mill's paper *On Liberty* to search for (hope- fully) new and refreshing theoretical tools to search for answers to the questions above.

In order to exemplify how the answers may not be so straightforward and optimistic as Mr. Zuckerberg expects, we can preliminarily devise the following observations respective to each question:

**Observation 1:** The relationship between State and People is constantly changing, just like any other social process or institution. Society is not immutable and the crystallization of social relations is a chimera dreamed by few and unfortunately attempted by many—fortunately, to no avail. What matters for us at this point is not whether society is changing, but why, how, and which old shapes will disappear and new relations emerge from such changes.

**Observation 2:** As will be demonstrated in this chapter and later in this book, the available data indicate that digital communication technologies and social media can and does have a profound impact on the levels of transparency by allowing the rapid distribution of information at very low cost, rapidly and to all areas of a country and sectors of society, as long as internet infrastructure is available. However, one note of caution comes at hand: the increase in the quantity of data available to the citizens may distract of confuse the people, deviating their attention from more pressing or relevant problems to less important issues that can catch more momentum. This questions overlaps partially with the next (Question 3).

**Observation 3:** The traditional structures and institutions of representation, such as parliaments, political parties, unions, civil society associations, guilds, the state bureaucracy, the press, among others, may lose importance as mediators between the government and the people, as the latter can access information and communicate directly with the former. This is particularly worrying as the editorial role such institutions have of selecting and organizing information, and devising rational strategies for attaining medium and long-term agendas may be lost.

Our study does not incur in the hubris of attempting to provide definite answers to the questions above, yet it is possible and useful to read Mill's work to extract theoretical tools helpful to better understanding each proposed problem.

## THE NEW RELATIONSHIP BETWEEN STATE AND CITIZENS IN THE DIGITAL COMMUNICATIONS AGE: A TALE OF TWO CITIES?

Until now, the patterns of relationship between the State and citizens have been different in democracies and dictatorships for virtually all measurable aspects. Digital media may cause the manners the State and the people interact in democracies and dictatorships to converge. This section will examine the possible convergences and differences in participation in democracies and dictatorships in how transparency mechanisms function in each type of polity. Decision-making procedures are perhaps the most striking difference between democracies and dictatorships, as in "Open Access Orders" (North et al. 2007) the Government is controlled by the society and is the main forum for lawmaking (Legislative branch) and conflict resolution (Judiciary branch); while in dictatorships, although may display the three branches, all formally exercising their proper functions, the actual decision making happens within the center of power, controlled by the political elites and protected from the control of the people, who have very limited, if any, access to the decision making process. Yet, all States must in one-way or another, find a minimum justification to support the exercise of power over society or in other words, obtain legitimacy. The previous chapter, also authored by me, discusses the concept and characteristics of legitimacy in more detail. Regarding the legitimacy of democracies and dictatorships, North et al. (2007: 9-10) explain:

*Limited access orders must carefully balance the distribution of elite interests within the dominant coalition. A shift in the incentives facing a major player that leads him to defect from the coalition and use violence or other means to forward his interests will produce instability, if not civil war [...] the internal structure of the dominant coalition is not robust. Any shock that changes the distribution of violence potential can force renegotiation of the distribution of rents; and actual violence is a constant possibility.*

Democracies enjoy what some scholars often call *input legitimacy,* i.e. support and recognition to the government by the governed based on principle and on the nature of the rules of the game, as the people will tend to support the actions of a government whose decisions are made with and by the participation of the people. Dictatorships cannot resort on this specific type of legitimacy and must obtain the obedience of the people by other means, especially the use of violence against the people. However, the use of illegitimate violence to obtain obedience does not provide legitimacy to the regime, and when such violence is frequently applied on the people, it harms any support or voluntary obedience that may exist more than build genuine support. For this reason, dictatorships often attempt to provide governance results, often in the form of economic growth, provision of welfare o any other actions that may convince the population that their government, albeit not democratic, governs to the interest of the people and to promote its wellbeing. The difficulty in this strategy to obtain legitimacy is that economic and social development which leads to improvements in the population's well being is often not easy to accomplish, especially in the scale continuity necessary for generating out- put legitimacy. Also, non-democratic regimes define policy by centralized decisions that tend to concentrate the benefits of governmental action in the hands of the governing elites (North et. al. 2007), which increases further the difficulty by dictatorships in obtaining legitimacy. This is the reason for the recur- rent resort, by dictatorships, to propaganda to convince the population that their government promotes their interest and well being, often distorting fact and creating untruthful versions of the reality that are imposed on the population.

Traditionally, institutions and social structures like political parties, unions, civil society associations, foundations, etc. have served as aggregators of preferences, collecting the aspirations of the people and condensing them into rationally devised programs or agendas to be pressed unto the government to shape public policy. In Democracies such actors tend to be numerous, as it would be expected of open access orders. In closed access orders, the channels for forwarding the people's desires to the government are more scarce and less efficient. In both cases though, such task has been done with less and less efficiency, leading such institutions, especially political parties to face a crisis of credibility (Cardoso, 2005).

This is particularly evident in the case of the authoritarian regime in China, where the Communist Party claims the monopoly of policy shaping and strategic planning for the whole society. In addition to that, the authoritarian government, describing itself as "servant of the people" (Bergruen & Gardels, 2013), claims sole responsibility in providing efficient governance. In spite of the spectacular economic growth the Beijing regime had obtained, which brought the Chinese GNP to the world's second position, above Japan and under the United States in only three decades, problems such as heavy pollution, massive migrations from rural to urban areas, a fragile land ownership regime and corruption of government officials, among others, (most of them resulting from the rapid economic growth itself) have hindered the output legitimacy that could have been created by the government's performance in making the GNP grow, according to Holbig & Gilley (2010):

*The contemporary politics of China reflect an ongoing effort by the ruling Chinese Communist Party (CCP) to reclaim the right to rule in light of the consequences of economic development, international pressures, and historical change. China stands out within the Asian region for the relative success the regime has achieved in that effort. While the CCP does face challenges to its legitimacy, those challenges are, for the most part, defeated by regime claims. In some respects, China is a classically Asian case: a democratic opposition struggles against the rational-legal and economic performance claims of the regime.*

In this context, social media has been extensively used by the Chinese people to obtain extra official information and exchange experiences, not only among themselves but to expose the reality of life in China to the international public opinion (MacKinnon, 2011). The government reacts, trying to pre- vent the spread of word by deleting messages and posts contrary to its interests in servers under its control (Qiang, 2011a), monitoring citizen's activities and eventually harassing them (Bergruen & Gardels, 2013; 68).

Perhaps the well known words by Lu Yuegang, a Chinese journalist in an open let- ter to the Beijing government, can provide a clearer description of the Chinese Communist Party *modus operandi* regarding information controls (Quiang, 2004):

*Your Excellency said that in order to guarantee the permanent position of power of CCP, we must depend on two weapons: guns and pens. We understand this. ...... But Your Excellency has forgotten, that is the political philosophy of a revolutionary party ...the logic behind this philosophy is not only to control the pen but to have this control backed by the gun... Whoever wants to question or to resist will be shown a gun.*

In the midst of this information war, with the people on one side trying to obtain and spread what they perceive to be true, and the regime using its power to broadcast a version favorable to its interests (and often managing to mislead the people), what contribution can political theory provide us? John Stuart Mill provides us to with a cornerstone principle on the validity of opinions that, for its intrinsic simplicity, may be valid for practical application both in democracies and dictatorships (Mill, 1989: 31 and 35):

*The opinion which is attempted to suppress by authority may possibly be true. Those who desire to suppress it, of course deny its truth; but they are not infallible. They have no authority to decide the question for all mankind, and exclude every other person from the means of judging. [...] Complete liberty of contradicting and disproving our opinion, is the very condition which justifies us in assuming its truth for purposes of action; and on no other terms can a being with human faculties have any rational assurance of being right.*

The confrontation of conflicting point of view puts at test the truthfulness of versions (or opinions, as Mill prefers), allowing the common citizen to choose between the version that seems more solid when under attack by other versions. Whoever, remains the problem of how the people can have access to such diverse versions on the facts, when the information is controlled by the government, like in the case of China? Too often in human history, authoritarian regimes successfully controlled all available sources of information to the people, who in spite knowing that the government's propaganda did not correspond to their direct experience, did not have any medium to produce, share, and obtain information to compare and confirm. This is in fact the case of China—The Chinese Communist Party, since the foundation of the regime, has held a monopoly on the exercise of political authority and has made information control a central part of its governing strategy, monitoring and controlling all vehicles of mass media (TV channels, radio stations, newspapers, magazines, etc.) to ensure the official version on the facts is the only avail- able version (Qiang, 2011b). In the age of social media and digital communications, all the efforts by the government (the Chinese government allocates more resources to information controls than to its military (Bergruen & Gardels, 2013)) may not be enough to keep the people in the dark, as the widespread of affordable personal communication technology may make it too hard to monitor and control too many users at the same time, all the time, and as users also learn how to technically circumvent the barriers posed by the government. According to MacKinnon (2011):

*Twitter is blocked in China. Still, a growing community is so determined to gain access to the widely used social-networking service and hold uncensored conversations with people around the world that these Chinese Internet users have acquired the technical skills to circumvent this censorship system—widely known as the "Great Firewall of China," a filtering system that blocks websites on domestic Internet connections.*

In this way, perhaps Mill's confrontation of different versions may in fact transform the relations ship between the State and the people in China, and other dictatorships alike, as the government's best efforts to contain the ever increasing number of dissident voices fail and it is forced to change its strategy towards the widespread use of social media and start monitoring digital interactions as a feedback tool to improve, now truly, the efficacy of governmental actions and obtain output legitimacy. This has been reported to be already happening in China and will be examined more closely below, in the discussion on new participatory patterns.

## WIDESPREAD OF INFORMATION TECHNOLOGY, TRANSPARENCY, AND ACCOUNTABILITY

Digital communication technologies seem to be everywhere: in every pocket, on every desk, at the tip of every finger there seems to be a internet-connected device, ready to link each one of us to all others in real time anytime. This scenario may have seemed unrealistic or impossible to an observer from less than a century ago, and yet it is supported by fact: The still timid penetration of internet us- age in the less developed areas of the globe is being compensated by the staggering growth observed in the past few years in Africa, the Middle East and Latin America. Indeed, those are not only the least developed areas of the world, but also where most dictatorships (Middle East and Africa) or young and most time fragile democracies (Latin America) can be found.

However, information technologies evolved to make the user less free. As computers were starting, they were accessible only to a few, but it was open for whoever could get their hands into one to change it, program it and make new things with it freely. As information technologies follow their path towards ubiquity by expanding their reach to virtually all persons in the developed world and a staggering growing number of people in developing countries, the technology became more sterile: devices come programmed and the user has very limited options to using it, let alone programming it beyond what the manufacturer has previously devised. As Zitrain (2008: 1-3) reports on the evolution of two of the most iconic digital communication devices of all times:

*This was not the first time Steve Jobs had launched a revolution. [...]Jobs, wearing his first suit, exhibited the Apple II personal computer to great buzz [...] The Apple II was a machine for hobbyists who did not want to fuss with soldering irons: all the ingredients for a functioning PC were provided in a convenient molded plastic case. [...] Some owners were inspired to program the machines themselves [...] The Apple II was a blank slate, a bold departure from previous technology that had been developed and marketed to perform specific tasks from the first day of its sale to the last day of its use. [...] Though these two inventions—iPhone and Apple II—were launched by the same man, the revolutions that they inaugurated are radically different. [...] The Apple II was quintessentially generative technology. It was a platform. It invited people to tinker with it. [...] The iPhone is the opposite. It is sterile. Rather than a platform that invites innovation, the iPhone comes preprogrammed. [...] Jobs was not shy about these restrictions baked into the iPhone.*

Interestingly, democracy followed a very similar pattern. In early democracies (what Mill calls "some of the popular governments of Greece" 1989:2), the people exercised power over themselves, according to the "last generation of European liberalism" (ibid:5). In those small cities where democracy was first devised and put to practice, as well as at the wake of modern era, a few representatives, elected by the people, exercised the government under very few limitations on their scope of action. This was so because the very vote by each citizen, that elected the representatives was believed to be enough to grant unlimited prerogative, or as Mill (ibid: 4-5) puts it:

*Table 1. World internet usage and population statistics June 30, 2012. Source: Internet World Stats: Usage and Population Statistics. Retrieved on January 20, 2014 from: http://www.internetworldstats.com/stats.htm.*

| World Regions | Population (2012 Est.) | Internet Users Dec. 31, 2000 | Internet Users Latest Data | Penetration (% Population) | Growth 2000-2012 | Users % of Table |
|---|---|---|---|---|---|---|
| **Africa** | 1,073,380,925 | 4,514,400 | 167,335,676 | 15.6% | 3,606.7% | 7.0% |
| **Asia** | 3,922,066,987 | 114,304,000 | 1,076,681,059 | 27.5% | 841.9% | 44.8% |
| **Europe** | 820,918,446 | 105,096,093 | 518,512,109 | 63.2% | 393.4% | 21.5% |
| **Middle East** | 223,608,203 | 3,284,800 | 90,000,455 | 40.2% | 2,639.9% | 3.7% |
| **North America** | 348,280,154 | 108,096,800 | 273,785,413 | 78.6% | 153.3% | 11.4% |
| **Latin America / Caribbean** | 593,688,638 | 18,068,919 | 254,915,745 | 42.9% | 1,310.8% | 10.6% |
| **Oceania / Australia** | 35,903,569 | 7,620,480 | 24,287,919 | 67.6% | 218.7% | 1.0% |
| **WORLD TOTAL** | **7,017,846,922** | **360,985,492** | **2,405,518,376** | **34.3%** | **566.4%** | **100.0%** |

*A time, however, came, in the progress of human affairs, when men ceased to think it a necessity of nature that their governors should be an independent power, opposed in interest to themselves. It appeared to them much better that the various magistrates of the State should be their tenants or delegates, revocable at their pleasure. [...] What was now wanted was, that the rulers should be identified with the people; that their interest and will should be the interest and will of the nation. The nation did not need to be protected against its own will. There was no fear of its tyrannizing [sic] over itself. Let the rulers be effectually responsible to it, promptly removable by it, and it could afford to trust them with power of which it could itself dictate the use to be made. Their power was but the nation's own power, concentrated, and in a form convenient for exercise.*

As the power could not be exercised directly by the ruled and an elite of representatives had to be elected to perform the daily functions of government, it soon became evident that even after being elected by the people, and under the close watch of public opinion, this elite of representatives was exactly that: an elite. And as such, the popular-elected government could still use power against the people. Of course, such representatives, or delegates mandates could always be "revocable ate [the people's] pleasure", but the notion that concentration of the exercise of power in a few hands, even elected hands, could still make democracies less and less democratic. Another problem noticed by Mill in the *government of the majority* is that the mere exercise of power by the people cannot guarantee that the decisions made are the wisest. Sometimes, the major- it may use its power in disastrous ways, or to persecute the portion of the people which does not share the dominant opinion. In Mill's words (ibid: 8-9):

*Society can and does execute its own man- dates: and if it issues wrong mandates instead of right, or any mandates at all in things with which it ought not to meddle, it practises a social tyranny more formidable than many kinds of political oppression, since, though not usually upheld by such extreme penalties, it leaves fewer means of escape, penetrating much more deeply into the details of life, and enslaving the soul itself. Protection, therefore, against the tyranny of the magistrate is not enough: there needs protection also against the tyranny of the prevailing opinion and feeling; against the tendency of society to impose, by other means than civil penalties, its own ideas and practices as rules of conduct on those who dissent from them; to fetter the development, and, if possible, prevent the formation, of any individuality not in harmony with its ways, and compel all characters to fashion themselves upon the model of its own. There is a limit to the legitimate interference of collective opinion with individual independence: and to find that limit, and maintain it against encroachment, is as indispensable to a good condition of human affairs, as protection against political despotism.*

This *Tyranny of the Majority* problem is by no means new to political science, and democratic regimes must provide guaranties and protections against ill-advised majority rule and oppression of the minorities by the majority. Such problems are tackled in con- temporary democracies with mechanisms such as the Rule of Law, and minimum rights and guarantee stone-written in the Constitution. However, if we step away from the more practical historical experience of democracies and look closer in a more purely logical level, what rests behind the "need" of limitations to the rule of the majority is a deeper feeling of distrust that the most of the people can make the best and wisest decisions consistently. Here perhaps I need a disclaimer: I do not believe or teach that minorities should be left alone under the rule of the majority, this in fact can and has in practice resulted in terrible and unacceptable oppression of the few by the many. Yet, when we move away from the history of democracies and look into the evolution of digital communications technologies, we observe that as such technologies became accessible to more and more people, until turning themselves into a mass phenomenon, the liberty to determine how each instrument could be used was gradually restricted to a small elite of programmers and moguls in the large corporations. In the words of Steve Jobs himself, at the occasion of the release of the first iPhone (New York Times, 2007):

*We define everything that is on the phone. You don't want your phone to be like a PC. The last thing you want is to have loaded three apps on your phone and then you go to make a call and it doesn't work anymore. These are more like iPods than they are like computers. These are devices that need to work, and you can't do that if you load any software on them. That doesn't mean there's not going to be software to buy that you can load on them coming from us. It doesn't mean we have to write it all, but it means it has to be more of a controlled environment.*

Thus, as the evolution of democracies from their direct participation form to today's mass democracies, made it more difficult to the common people to actively and directly participate in the government, information technology became less inclusive and less participative as the use-base broadened. As we have seen, this process is not new to political scientists, especially not to John Stuart Mill. As governments (democracies and dictatorships alike) try to incorporate digital communications technologies in the exercise of governmental functions, a number of opportunities for improvement in governmental transparency and inclusiveness are (Bertot et. al., 2012):

- *Democratic participation and engagement, using social media technologies to engage the public in government fostering par- ticipatory dialogue and providing a voice in discussions of policy development and implementation.*
- *Co-production, in which governments and the public jointly develop, design, and de-liver government services to improve ser- vice quality, delivery, and responsiveness.*
- *Crowdsourcing solutions and innovations, seeking innovation through public knowledge and talent to develop innovative solutions to large scale societal issues. To facilitate crowdsourcing, the government shares data and other inputs so that the public has a foundational base on which to innovate.*

These broadened opportunities for participation in government, co-production of policy and services and problem-solving are a clear improvement in participation levels for both contemporary mass-democracies, where the people have been alienated from the exercise of governmental functions as in dictatorships where the people never had the opportunity o participate. However it is symptomatic that this increased inclusiveness and broader opportunities for participation may be sought after even by dictatorships, which may wish to use digital communication technologies in a manner that allows the government quicker and more precise feedback on the acceptance of policies that would otherwise be imposed by the governing elite on the people. The Chinese Communist Party, as an example, uses the myriad of social commentary and opinion produced and posted online for two purposes: (1) to detect and control critical and *dissident* voices, and (2) to collect feedback from the people on the popularity and efficacy of policy, an extremely useful aid in reshaping and fine-tuning governmental action in the Chinese performance-legitimacy based model (Berggruen & Gardels, 2013:76).

All in all, the use of information technology raises questions regarding privacy of citizens, security, transparency and corruption in all levels of government. Information technology in government may take two primary directions: (1) top-down, where the initiative to communicate is taken by the government, which uses its servers to provide services and/ or information, such as application to services (such as wealth care programs), or the avail- ability of vital records; and (2) a bottom-up movement, where the people use social media to obtain and share often scarce information on any matter of the public's interest and coordinate and organize movements and responses to the government's actions and policies. Regardless the direction of these movements, digital communication technologies are increasingly present in government in democracies and dictatorships alike.

## TRANSPARENCY, CORRUPTION, AND ACCESS TO SERVICES: THE IMPACT OF SOCIAL MEDIA

Another form of people's participation in government is the monitoring of governmental actions. Mill points out that it has been the tradition in History that the people do not trust their rulers, or that the people expect the rulers to exercise their power to ends that do no meet (except occasionally and coincidentally) the interests of the people:

*The rulers were conceived (except in some of the popular governments of Greece) a in a necessarily antagonistic position to the people whom they ruled. They consisted of a governing One, or a governing tribe or caste who derived their authority from inheritance or conquest, who, at all events, did not hold it at the pleasure of the governed. (Mill, 1989:2)*

Far from being an ancient problem exclusive to the Greek city-states, the people's distrust of their rulers call for constant watch over the actions and policies adopted by the government on society. In fact, the past three decades have seen demand for transparency in government as a tool to combat corruption in both public and private spheres to go from taboo to a kind of global subject-of-the-day (Transparency International, n.d.a). In the fight against corruption, Bertot et. al. (2010) say:

*Many governments envision the use of ICTs as a means to promote efficiency and transparency at the same time . ICTs in general show promise as an effective means of reducing corruption, but social attitudes can decrease the effectiveness of ICTs as an anti-corruption tool . Case studies and statistical analyses indicate that ICTs hold a great deal of potential for – and are already demonstrating benefits in – anti-corruption, particularly by enhancing the effectiveness of internal and managerial control over corrupt behaviors and by promoting government accountability and transparency.*

Traditionally, the printed press has taken the task of providing, in democracies, the people with information regarding the actions of government and the reality of social life independently from the official propaganda. Liberty of expression (and of thought and of discussion) has indeed been considered one of the most important and fundamental guarantees against the abuse of power and tyranny. Mill says that " the 'liberty of the press' [is] one of the securities against corrupt or tyrannical government" (1989:28). In the text, the freedom of press is equal to freedom of thought (or opinion) and freedom of discussion. This is so because in Mill's argument they are identical freedoms and the same right: that of having an opinion and expressing it publicly to the common debate of society. For that reason, the freedom of press, far from having its semantic core in the term "press", or the traditional journalistic or editorial bodies, actually means freedom to publish one's opinion, to make oneself's views and thoughts known to the greater public.

Until the advent and widespread of digital communications and social media, broadcasting and publishing was an expensive business, and the exercise of the "freedom of press" was restricted to the few who had access to the increasingly elaborate and costly instruments for broadcasting and opinion o even a finding on secret governmental activity, except to a rather restricted public, often encompassing one's family and friends. In the age of social media, however, every citizen, each individual has the potential power to reach thousands and perhaps millions of people in a fast, accessible and affordable process.

One common aspiration of all peoples, governed by democracy of under a dictatorship is that their government is effective in promoting the conditions to security, well-being and prosperity for all persons living I its territory. In fact, this governmental efficacy is one of the bases for legitimacy, as the previous chapter has exposed. Transparency International's (n.d.) definition of corruption, as "the abuse of entrusted power for private gain. It hurts everyone whose life, livelihood or happiness depends on the integrity of people in a position of authority" in direct conflict with this fundamental source of legitimacy and often *raison d'être* on many regimes. Fighting corruption is an even more pressing necessity in developing countries, where resources are comparatively scarce and needs more numerous. In fact, developing countries which do not fight or control corruption in government and the private sector may suffer great loss of reputation and even economic advantages and aid coming from developed nations. Not only aid, but aldso direct productive investment. Ramos (2015) notes that aid and foreign investment in less developed nations correlates to increased levels of corruption, while in wealthier, more developed parts of the world resources and investment allocations tend to be more transparent. It is only reasonable to infer that foreign private investment may tend to prefer ore honest and transparent economic environments.

The affordability of social media tools for transparency makes digital communications technologies even more necessary for the peoples in the poorer nations to attempt to curb corruption and force a more transparent and honest governance by its rulers, as transparency seems to be an effective tool to fight corruption (ABC, 2001:87; Brett et. al. 2010). The following chapters in this book report on a wide diversity of international experiences where social media and information technologies have been used by governments and the people to promote transparency, curb corruption, and improve general living standards by better governance. In fact, social media has been used as an instrument to improve the quality of governance in many areas and countries, as Brett (2012) states:

*Much government activity is now focused on social media, with social media becoming a central component of e-government in a very short period of time U.S. federal agencies have been using blogs, microblogs, wikis, and social networking sites, and even virtual worlds, among other social media, to create records, disseminate information, communicate with the public and between agencies for several years. The General Services Administration has even created a standard agreement for social media providers to allow for government usage of social media services. The Obama administration has made a priority of the use of social media technologies, and the new Federal Chief Information Officer is strongly encouraging the expansion of these activities.*

*Not only is the Obama administration strongly encouraging agencies to use social media to provide information, communicate with members of the public, and distribute services, it has also made a priority of public usage of social media to participate in government. And many members of the public already expect that government services will be available electronically and that government agencies will be accessible via social media technologies.*

We must also note that not only people around the globe are using social media to demand better governance from their rulers, but also governments have resorted to digital communications to provide access to a wide variety of services and information to their populations, in democracies and dictator- ships alike, such as vital records, tax refund reports, health care, visas, etc., although the regulatory framework in many countries still seems insufficient to solve concerns about privacy, data management, social inclusion, and accessibility (Bertot, 2012):

*Social media technologies have already taken an important place among means of communication between government and members of the public and are poised to continue to take on greater prominence as a mechanism of government information and services are more content moves onto this platform and more users come to expect to use social media as a primary method to interact with government. At its best, social media has the potential to simultaneously make government more reachable, available, and relevant to users, while offering users more opportunities to become actively engaged in government. To ensure that social media technologies enable some or all of this lofty potential, the policy issues related to social media must be closely examined and addressed while the uses of the technologies are still developing and evolving.*

It is thus necessary to devise not only new regulatory frameworks to discipline the actions of public agencies to protect the privacy and security of the people they serve, without undercutting the agencies effectiveness. Such regulations must also promote maximum liberty to citizens to seek better services and information from governmental agencies, as well as expressing and sharing their opinions, without restricting the liberties of other citizens or groups, for in Mill's words:

*[...] the sole end for which mankind are warranted, individually or collectively, in interfering with the liberty of any of their number, is self-protection. That is the only purpose for which power can be rightfully exercised over any member of a civilized community against his will, is to prevent harm to others.*

The relationship between digital telecom Technologies, social media and democracy has taken a startling character in a young, significant and consolidating democracy: Brazil. Social media and digital communications were instrumental and decisive in the 2018 presidential elections where Jair Bolsonaro was conducted to the Presidency with impressive vote count. Bolsonaro, an old-school, typically pork-seeking politician, who was based in metropolitan Brazil, close to the elites and middle-class voters of Rio de Janeiro, São Paulo and Brasilia lacked proximity to the hinterlands and impoverished suburbs where much-needed voters lived. In addition to that, due to specific electoral law rules in Brazil, Bolsonaro had very limited access to officially assigned TV advertising and relied heavily on social media campaigning in platforms such as WhatsApp, Facebook and Tweeter, much more that any of his opposing incumbents. It worked.

Moura and Coberllini (2018) note that smartphones are the most often used and decisive technological device used by voters in contemporary democratic Brazil, in line with other mass movements worldwide, such as the Arab Spring, Occupy Wall Street and more recent Hong Kong yellow umbrella protests. Also, social media campaigning and viralising impacted decisively the outcomes of presidential elections in neighboring Colombia and Mexico (both in 2018), and left political analysts, polling experts and traditional journalists and politicians who still trusted conventional campaigning and mass media channels, such as TV, the printed press and the radio in unabashed awe and disarray. In fact, Bolsonaro's adversaries in the political elite journalistic establishment, who calculated that conventional campaigning on TV radio and newspapers would be sufficient to deconstruct his public image and reputation unto voters were left to watch helplessly as popular support to the outsider, propelled by social media, grew and strengthened.

Perhaps the silver lining in this whole digital disruption is the increased empowerment of the common folk to express their political opinions and preferences and broadcast them in ever increasing circles. Yet, while the common citizen's access to information and also to the means to share and publicize their opinions and views increase and related costs plummet, one problem arises: that the fact-checking and editorial activities usually done by professional journalists and content editors that grant traditional news media are less and less significant vis-à-vis the volume and variety of information that reaches the common citizen through social media. Indeed, the proliferation of fake news on digital and social media poses a real threat to the trustworthiness of publicly available information in democracies worldwide. Content editing, fact-checking and other mediating activities do help, yet increased controls on free expression of opinions may restrict and even pose a threat to what according to Mill (2017:138) is a fundamental element of any free society. Freedom of expression cannot harbor false information, exposure of privacy, disrespectful and offensive discourse, or reputation bashing. A 2018 report on trustworthiness of brands, traditional and social media by Edelman (2018) notes decreasing levels of trust among citizens in democracies and authoritarian regimes alike:

The scenario in the other large democracy of the Americas, the United States, is significantly similar. Wolf (2018) reports that Donald Trump's ascent to the White House was tinted by a traditional-media-bashing rhetoric, through social media and resulted quite efficient.

Increasing discredit of traditional media as well as new social media may leave the common citizen very limited access to trustworthy information and hinder everyone's ability to check facts and form informed opinions and decisions. Indeed traditional communications and political representative institutions such as parliaments, political parties, unions, civil society associations, state bureaucrats and private technocrats role in mediating the relationship between the State and the governed. We may

*Figure 1. Traditional Media, Search Trusted; Social Media Not. Source: Edelmand Trust Barometer. Special Report: Brands and Social Media. Retrieved on 14 July, 2019 from: https://www.edelman.com/sites/g/files/aatuss191/files/2018-10/2018_Trust_Barometer_Brands_Social_Media_Special_Full_Report.pdf*

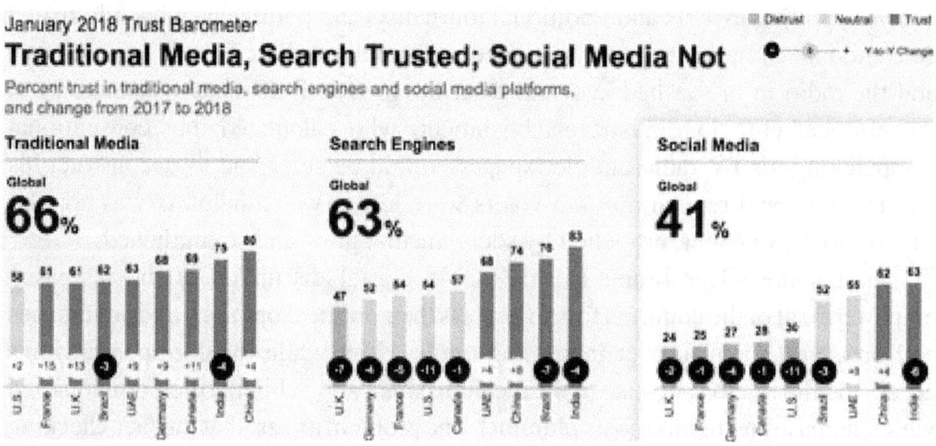

very well be moving towards an era where self-restraint and direct peer-reprobation become the main means of socially-accepted behavior control. And, as Mill insisted:

*A person who shows rashness, obstinacy, self conceit—who cannot live within moderate means—who cannot restrain himself from hurtful indulgences—who pursues animal pleasures at the expense of those of feeling and intellect—must expect to be lowered in the opinion of others, and to have a less share of their favourable sentiments; but of this he has no right to complain, unless he has merited their favour by special excellence in his social relations, and has thus established a title to their good offices, which is not affected by his demerits towards himself*

We only worry that, in the mass democracies of the 21st century self restraint and peer controls are not enough to keep our democracies functioning.

## CONCLUSION

The widespread use of social media by citizens and governments has had a profound and yet not fully understood impact on the relationship between those same governments and citizens. The traditional categories of political science such as democracy, transparency, freedom of expression, participation, representation, etc. all may aid in the comprehension of such changes, although those very concepts may need to be re-visited and re-defined to better serve to interpret and understand the transformations that are now taking place. In this chapter I attempted using some very classical formulations on the rules and limits of interactions between the State and the people from John Stuart Mill's *On Liberty* to provide parameters to understand changes that are taking place in such a dramatic pace that perhaps only the stability and sturdiness of well established doctrine can provide.

Our first concern referred to the nature of changes in the relationship between governments and their peoples. In both democracies and Dictatorships the relationship between the State and Citizens is being impacted by broadened possibilities of citizens obtaining information and expressing their concerns. Governments are also using social media and other expressions by the citizens to shape policy more effectively, thus improving the quality of governmental services. This effect may have both positive and negative consequences. One is that citizens may become better served, and life quality may improve in the areas affected by certain policies. One negative consequence is that dictatorships may use the immediate popular feedback to address some of their people's concerns more effectively to obtain legitimacy and delay popular discontent that could weaken the regime.

Regarding the impacts that new digital communication technologies have in the transparency and accountability of governments, our theoretical inquiry indicates that the new digital communication technologies may not only increase both citizen's and governments capacity to obtain information and participate in devising policy, but also allow for more opportunities for the people to monitor governmental activity and exercise their traditional "people's distrust of the rulers". Improved transparency tends to lead to less corruption, or diminishing misuse of resources that were intended for promoting the common good. Also, social media may be used by citizens to organize protests and aid in keeping undemocratic government accountable, at least to some extent. Yet, some concerns regarding the collection of data from citizens and the management of such data arise: devising new regulatory frameworks for privacy, data management and security of information has come up as one necessary measure to aid in alleviating such concerns. Perhaps our third question is where we can provide the least definite answers at this point: How are the representation system, and the State bureaucracy affected by the increased participation by the people in the shaping of policy and legal decision-making? In fact, one can easily defend a relative decay in the importance of traditional representative and State/People mediating bodies such as parliaments, unions, and civil society associations. As the people are empowered by new technologies to influence directly the shaping of policy and steering more effectively their government's decisions, such mediators may start being deemed as unnecessary and wasteful of limited societal resources. As true as that in fact is, the mentioned traditional representative bodies also have an "editorial" role in collecting often disarticulated and fragmented aspirations in society and transforming them into a more comprehensible, strategic and rational policy agenda, and delivering it to the government. More direct participation may also increase the difficulty in implementing long-term more rational and strategic policies for the development and common-good of the societies involved.

In a nutshell, no one doubts, at this point, that the widespread of social media and digital communications will have and is having deep and unavoidable impacts on the relationship between the State and citizens. However, such impacts raise both hopes of a better future for democracies and dictatorships as well as concerns on privacy, freedom of speech and thought, consistency of public policy, quality of governmental services, and even the legitimacy and lifetime of regimes. The theoretical tools of classical political science may, however aid us in better understanding such processes and steering change for the best.

## REFERENCES


Bac, M. (2001). Corruption, Connections and Transparency: Does a Better Screen Imply a Better Scene? *Public Choice*, *107*(1-2), 87–96. doi:10.1023/A:1010349907813

Berggruen, N., & Gardels, N. (2013). *Intelligent Governance for the 21st Century—A Middle Way between West and East.* Polity Press.

Bertot, J. C., Jaeger, P. T., & Grimes, J. M. (2010). Using ICTs to Create a Culture of Transparency: e-Government and Social Media as Openness and Anti-Corruption Tools for Societies. *Government Information Quarterly*, *27*(3), 264–271. doi:10.1016/j.giq.2010.03.001

Bertot, J. C., Jaeger, P. T., & Hansen, D. (2012). The Impact of Polices on Government Social Media Usage: Issues, Challenges, and Recommendations. *Government Information Quarterly, 29*, 30–40. doi:.giq.2011.04.004 doi:10.1016/j

Cardoso, F. H. (2005, Sept.). Political Parties. *Foreign Policy*, 41.

Holbig, H., & Gilley, B. (2010). Reclaiming Legitimacy in China. *Politics & Policy*, *38*(3), 395–422. doi:10.1111/j.1747-1346.2010.00241.x

Internet World Stats. (n.d.). *World Internet Usage and Population Statistics June 30, 2012*. Retrieved on January 20, 2014 from: http://www.internetworldstats.com/stats.htm

MacKinnon, R. (2011). China's "Networked Authoritarianism". *Journal of Democracy*, *22*(2), 32–46. doi:10.1353/jod.2011.0033

Mill, J. S. (1989). *On Liberty*. London: The Walter Scott Publishing Co.

Moura & Corbellini. (2019). A eleição disruptiva: por que Bolsonaro venceu. Rio de Janeiro: Record.

North, D. C., Wallis, J. J., Webb, S. B., & Weingast, B. R. (2007). *Limited Access Orders in the Developing World: A New Approach to the Problems of Development.* The World Bank Independent Evaluation Group. doi:10.1596/1813-9450-4359

Qiang, X. (2011a). The Battle for the Chinese Internet. *Journal of Democracy*, *22*(2), 47–61. doi:10.1353/jod.2011.0020

Qiang, X. (2011b). The Rise of Online Public Opinion and its Political Impact. In S. L. Shirk (Ed.), *Changing Media, Changing China*. Oxford, UK: Oxford University Press.

Quiang, X. (2004). A Bold new voice – Lu Yuegang's Extraordinary Open Letter to Authorities. *China Digital Times.* Retrieved on March 27, 2014 from https://chinadigitaltimes.net/2004/07/a-bold-new-voice-lu-yuegangs-extraordinary-open-letter-to-authorities/

Ramos. (2015). Corrupción y ayuda al desarrollo. Evidencias, teoría y aplicaciones para España. Madrid: Universidad CEU-San Pablo.

The New York Times. (2007). *Phone Shows Ap- ple's Impact of Consumer Products.* Retrieved on December 19, 2013 from: http://www. nytimes.com/2007/01/11/technology/11cnd- apple.html?pagewanted=print&_r=0

Transparency International. (n.d.a). *Our History.* Retrieved on April 11, 2014 from http:// www.transparency.org/whoweare/history

Transparency International. (n.d.b). *Government Defence Anti-corruption Index – Methodology (Long).* Retrieved on April 11, 2014 from http://government.defenceindex.org/ sites/default/files/documents/GI-methodology-long.pdf

Wolff. (2018). Fogo e fúria: por dentro da Casa Branca de Trump. São Paulo: Objetiva.

Zittrain, J. L. (2008). *The Future of the Internet-- And How to Stop It.* New Haven, CT: Yale University Press & Penguin.

Zuckerberg, M. (2012). *Letter from Mark Zuckerberg.* Washington, DC: The United States Securities and Exchange Commission. Retrieved on February 26, 2014 from http://www.sec.gov/Archives/ed- gar/data/1326801/000119312512034517/d287954ds1.htm#toc287954_10

## KEY TERMS AND DEFINITIONS

**Classical Democracy:** Roughly a type of political regime that developed in some cities of Greece (esp. Athens) circa the fifth century BC and that consisted in all citizens gathering to discuss propose and decide policies, rules and actions in the government of the city.

**Digital or E-Democracy:** A relatively new and still loosely defined concept, refers to the use of modern digital communications technologies in the interactions between the citizens, civil society, and the business sec- tor with governmental agencies to inform, shape, help implement and evaluate policy, as well as to monitor the actions and results of government.

**Direct Democracy:** A form of government where the citizens are called to participate directly in every discussion and decision. Traditionally, direct democracy is only possible in relatively small communities where frequently gathering all citizens is not so cumbersome. Still, most representative democracies allow for some forms of direct democratic procedures such as referendums and plebiscites on occasion. Digital communications technologies are held as a promising new method to increasingly facilitate direct democracy in modern mass democracies.

**Liberty (Mill's):** For the narrow purposes of this chapter, liberty can be defined as the condition in which an individual is restricted in his actions only by the possibility of harming others. This results in a very broad scope of action to societal actors, and a rather narrow one for the government, which can act only to protect the liberty of all individuals and guarantee that no harm is inflicted on any one of them.

**Representative Democracy:** The most common solution to the problem of frequently gathering all citizens for all discussions and decisions of direct democracies, as the population and territory of modern nation-states increased, representative democracy provides that a few citizens be chosen by the vote to all other citizens to represent the interests and aspirations of the political community in the government.

**Tyranny (Modern):** Unrestricted government, which can act freely according to its own will and independently from the consent and interests of the governed. In general libertarian theory, the freer the government, the less free the citizens, or in other words, tyrant government's freedom of action result in citizens restricted in their freedom.

**Will of the Nation:** One rather undefined term in Mill's writing, refers to the compound of projects, desires, aspirations and objectives of a relevant national community. Determining what the will of the nation is in any real context is the business of politics to devise the actions of government, public policy and the rules that govern society's life and the actions of the individuals. In democracies, the will of the nation is the result of the deliberations of the citizens, either directly or through their representatives.

# Section 2
# Case Studies: The Americas

# Chapter 3
# The Mexican Spring:
## Two Experiences on State-Society Relations in Mexico

**Atzimba Baltazar Macias**
*Independent Researcher, Mexico*

## ABSTRACT

*The chapter aims at understanding a recent phenomenon in Mexican politics: the use of internet and social media as a new and powerful resource for mobilization and social participation in the policy process. Based on a review of two relatively recent movements in Mexico (#YoSoy132 and The Wirikuta Defense Front), the chapter argues that although the internet is still restricted to the middle and upper classes, the use of social media and its impact transcends class boundaries, draws public attention, creates a valuable social capital for mobilization, and influences the decision-making process. The chapter does not intend to provide evidence to the theoretical discussion on why and how social media enhances political participation and mobilization; rather, it reflects on the features shared by these two movements in order to draw some lines for further research. It finds that, if used appropriately, social media is actually an effective tool to facilitate mobilization and modify the public agenda.*

## INTRODUCTION

The chapter aims at understanding a recent phenomenon in Mexican politics: the use of the Internet and social media as a new and powerful resource for mobilization and social participation in the policy process. The paper argues that although the

DOI: 10.4018/978-1-7998-3152-5.ch003

Internet is still restricted to the middle and upper classes, the use of information and its impact transcends class boundaries, draws public attention, creates valuable social capital for mobilization and influences the decision-making process.

The chapter is divided into three sections. In the first section, I will review the academic discussion on the use of Internet and social networks. The second section will analyze two major movements that have arisen in Mexico through social networks: #YoSoy132 and the campaign to protect the *Wirikuta* area. Finally, the chapter will discuss the use of social networks for mobilization and participation in the consolidation of democracy in Mexico.

## THE USE OF INFORMATION TECHNOLOGIES FOR MOBILIZATION AND PUBLIC PARTICIPATION

This section provides a brief review of the academic discussion on the impact of information technologies (ITs) for enhancing participation and mobilization. Although the chapter is not intended to contribute to the theoretical discussion on the use of information technologies, it will serve to frame the analysis of the Mexican case discussed in the following section.

There is a common acceptance that the Internet, in general, and social media, in particular, has incorporated a new feature in the way people acquire information that has triggered the monopoly that traditional media have had. There is also an agreement that the web socializes information at such a high speed unimaginable a few decades ago. From this perspective, the impact of the Internet and ITs is not minor: they have disempowered the State and traditional media since they cannot longer regulate public transaction and monopolize information (Turnsek and Jankowski, 2008). This makes political information more attainable to a larger share of the population (Stanley and Weare, 2004).

However, the extent to which the Internet and social media enhance political participation and mobilization is more debatable. There is still little evidence to support this argument. On the one hand, some argue that although ITs help in socializing events that happen around the world in few seconds, the use of ITs have an impact only on those who are already politically involved in a specific cause by providing more information and disseminating it. Also, they argue that, since Internet use is still limited to upper and middle classes, it has not yet replaced the traditional institutions for mobilization, such as parties or unions; neither it has replaced traditional media as the most important tool to gather and disseminate information. In addition, social media have been ripe for the proliferation of fake news –information presented as news that is low in facticity–, since it has lifted many obstacles to producing and distributing news that were present in the previous mass media age. Their impact is higher, in terms of political preferences, as they are aimed at specific constituencies (with the help of big data and algorithms) who trust most of the content in their social media feed (Owen, 2017).

On the other hand, those who argue that Internet has a positive impact on political participation and mobilization state that Internet and social media allow people to gather information, express their opinion and mobilize; in their view, having more information from direct sources convinces people to join a specific cause because they feel that their participation is bound to other activists who share their experiences and achievements. In other words, people's access to social media networks has a positive effect on their sense of political efficacy and tendency to participate in politics (Gil de Zuniga, et al., 2010).

In this sense, Internet has a transformative effect in society in at least three ways: it opens traditionally inaccessible information to public scrutiny; it opens spaces for unmediated deliberation in which people can interact with one another, with different communities and with people from another country; it changes the way representatives perform because now they can be questioned directly and this, in turn, changes the way in which traditional powers, unions, parties and media relate to their constituencies and publics (Coleman, 2001).

In this view, civic and political mobilization is an outcome of individual agency and mobilizing agency (Norris, 2002). On the individual agency, mobilization depends on economic resources, motivation and skills. On the mobilizing agency, political activity relates to the structures for spreading information and for the reasons why individuals mobilize in the first place. Social media has specific and new characteristics that affect these two dimensions of political and social mobilizations:

*The networks functionalities and affordances characterizing social media makes them potentially very efficient in spreading information about upcoming events and demonstrations within a population. The small world effect makes it likely that the same information will reach all parts of a network, and therefore reach differentiated sets of groups of individuals. One may also argue that information cascades may have a not only an informational, but also a motivational effect, since social media allow making visible the choice of that the others make to join a group or to sign up for an event (Bernard, Steen-Johnsen and Wollebaek, 2012).*

Based on these assumptions, it might be argued that political mobilization in the Internet-based day is broader and easier than in earlier mobilization processes.

However, based on the evidence available and on the scope of studies, certain characteristics must be in place for Internet and social media to actually enhance social and political participation. Some argue that the most significant variable in explaining how the use of social media has a positive impact on political participation and mobilization is the political regime: in democracies, the organization of non-violent protest constitutes a central feature of the regime, since it is a legitimate resource for giving voice to dissidents. Under authoritarian regimes, on the contrary, civil protest is more difficult because ordinary citizens do not control personnel, information and do not have political resources (Breuer, 2012).

*[...] While material and moral grievances may provide an abundance of motives under authoritarian rule, the ability of civil society to funnel these motives into collective action is usually thwarted by the fact that the public sphere is sealed: the national narrative is controlled by a government which usually resorts to a mix of censorship intimidation and persecution to suppress information on economic malperformance, human rights violations, corruption or any other issue which may negatively reflect on itself [...] By providing a space for free speech, the Internet poses an existential threat to the ability of authoritarian governments to control this national narrative. Web-based communication technologies make it possible to expose non-governmental narratives to a wide public (sic) (Breuer, 2012, pp. 5-6).*

This dissemination of information reinforces people's desire to protest and promotes collective action at the macro-level. By sharing information on past protest events, citizen's hope on the outcome also rises and enhances collective action (Breuer, 2012).

This argument can lead to the conclusion that the use of Internet and social media can only lead to social mobilization in authoritarian or semi-democratic regimes (Tunisia, Egypt, Mexico and Chile); however, it does not explain other mobilizations that have taken place in democratic regimes such as the 15-M movement in Spain or the environmental movements in Europe. There is an interesting study that found that in the Norwegian context, there are significant differences between those mobilized by traditional established civil society and those mobilized through social media: social media seems to represent a channel that supplements established political and civil society organizations by reaching different and less privileged groups (Bernard, Steen-Johnsen and Wollebaek, 2012). However, there is not sufficient evidence to make this argument conclusive: Norway is an old democracy and might be an outlier when compared to other realities worldwide.

A more plausible explanation considers what use people give to social media. It can be used for different purposes: people can use social media for information gathering, as news source; they can also use it as a space for expressing political opinions and, finally, it can be used as a venue for joining a cause and finding mobilizing information (Valenzuela, 2011). Interestingly, Valenzuela (from a survey of a random sample of adults living in urban areas in Chile during 2011) finds that the frequent use of social media, regardless of the purpose, was positively and significantly related the Chilean 2011 demonstrations (Valenzuela, 2011).

Other studies relate the type and use of social media with the likelihood of social mobilization. While the Internet increases the speed, the scope and widens the sources of information, only web 2.0 platforms allow people to express their opinion and to promote collective action. Unlike web 1.0 that does not allow for interaction between the transmitter and the receiver of information, social media platforms like Twitter, Facebook, YouTube Google+, Vimeo and blogs represent an important tool for promoting debate; they offer possibilities for the citizens to express their opinion to their leaders and other political actors and they allow for sharing information. However, Gladwell (2010) claims that the Internet only fosters weak ties that do not enable high-risk activities like protesting. Furthermore, messages uploaded to web 2.0 platforms can be replicated with no additional effort to other audiences (González, 2012). Social media is, thus, "a mass conversation among private actors that takes place in a public space" (Vega and Merino, 2011). From this point of view, political mobilization can only be enhanced by web 2.0 platforms.

The extent to which the use of social media favors political mobilization is intrinsically linked to certain aspects of the movement. First, the message published in social media must be clear enough to engage people to join; second, the movement should allocate its advocacy strategy in the appropriate media: two actors with the same propensity might be recruited at different times if one is connected to a larger number of people; third, messengers must be connected to the overall or the main network so that they can translate and disseminate the message to a wider audience; fourth, exposure to multiple sources can be more important than multiple exposures, thus, the movement has to insert within a widespread scope of tools and networks so that it secures that the message will be delivered to the people that are likely to engage in mobilization (Bailón, Borge-Holthoefer, Rivero and Moreno, 2011).

Social media also exacerbate trends that undermine the ideals of free speech and democratic life through the dissemination of fake news and the polarization of users around public issues. The accessibility of social media allows any user or interest group to spread information easily without third party filtering, fact-checking or editorial judgment. Therefore, authoritarian governments, political parties and interest groups may disseminate false information and provoke public outrage in a matter of minutes with limited repercussions. During the US presidential race in 2016, Donald Trump's campaign team hired Cambridge Analytica to target voters with specific profiles through the propagation of false news about Hillary Clinton. In addition, interest groups, such as the National Rifle Association, may use social media to spread distorted facts and perspectives in order to divide public opinion and advance their agendas. This has engendered longstanding debates regarding the negative effects of social media for political participation and social mobilization, although proposals aim more at regulating content in social media than banning them altogether.

From this very brief review of the academic discussion on the use of Internet and social media, it is evident that there is still room for research, however, as I will argue in the following sections, the more the digital gap widens, the more the use of social media will impact the relationship between society and the State. First, citizens are being empowered by Internet and social media, for they are no longer subject to traditional media to be informed, now they can gather information from unofficial sources from all over the world. Second, traditional institutions (parties, governments, media) are forced to take into account public social media opinion, redesign communication channels and widen participation tools, the risk of which lies in the dissemination of fake news to divide public opinion and advance specific interests. Third, activist, non-governmental organizations and other groups are starting to perceive social media as a useful and relatively cheap tool to bring their issues into the political agenda and to organize and mobilize their supporters. Finally, access to Internet is becoming more and more common among different social groups; thus, access to information is becoming more democratic. The Digital era is here and it is becoming more and more difficult to stay out of it; this is especially true when it comes to students and for people who are in their twenties; traditionally the ones that are more likely to mobilize.

## THE MEXICAN CASE

Mexico is a recent democracy, the Institutional Revolutionary Party (Partido Revolucionario Institucional, PRI) ruled for 70 years and it was not until 2000 that the National Action Party (Partido Acción Nacional, PAN) came into power through democratic elections. On the economic arena, Mexico is a developing country, which bases its economy on agricultural, manufacturing and oil exports. In 2015, Mexico had a population of 119, 530, 753 (INEGI, 2015) out of which 45% lived in poverty. According to the World Bank, in 2018, 66% of the population had access to Internet. In recent years, Mexico has dealt with high rates of unemployment and the rise of violence rates associated with drug trafficking. Although macroeconomic indicators are acceptable, Mexico is one of the most unequal countries worldwide.

Within this context, this section will review the evolution of two important movements through social media: #YoSoy132 and the defense of *Wirikuta*, a sacred ceremonial area of the *Huichol* Indians in San Luis Potosí. It should be noted that the selection of cases is not random; I try to find similarities and differences in the use of social networks of two movements that are different in nature and how they managed to gain public attention and influence the decision-making process.

## #YoSoy132

2012 was an electoral year and after twelve years where the right-wing National Action Party ruled, polls positioned the candidate of the Revolutionary Institutional Party as the virtual winner of the federal elections. This was due in part to the social perception that PAN had lost control of national security because of its policy against organized crime; and because the Democratic Revolution Party (Partido de la Revolución Democrática, PRD), the main left-wing party at the time, had experienced an internal division.

The common feeling was that the presidential elections would be far from competitive and with a predictable result. In this context, on May 11, 2012, Enrique Peña Nieto, the candidate from the PRI, visited the Universidad Iberoamericana, an upper-class university in Mexico City. At first, the lecture followed the common trend of this kind of events: supporters cheered the candidate and students against him chanted slogans and unfurled posters. It was the last intervention of the candidate that ignited the temper of the attendants: a question related to his role when he was governor of the state of Mexico and he ordered the containment of the uprising in San Salvador Atenco in 2006, which resulted in the arrest of 207 people and the death of two young men, in addition to reported cases of authority abuse, police brutality and violations of women. The candidate's response argued that his conduct was constrained within legal boundaries in order to preserve the public peace and he denied that the authorities had committed abuses. This statement provoked boos from the crowd (González, 2012). After his intervention, the candidate intended to give an interview to the university radio station, but he was forced to take refuge in the college bathrooms because students blocked him while shouting slogans. This forced his security team to make the decision to leave the University for not compromising the candidate's safety or jeopardize him.

Students' unrest came the next day when newspapers reported that the candidate's visit to the university was a success, despite a boycott orchestrated by people outside the university. These allegations, coupled with what seemed to be biased coverage by the media, made the students who had been present at the previous day's event express their discontent. During that weekend, the students created a Facebook event, which they called "Video for truth" inviting students who had been at the event to send a video identifying themselves, with their IDs in hand, as students from the Universidad Iberoamericana. 131 students responded to the call and uploaded a video on YouTube called "131 students from the Ibero respond", in which they started with images from media coverage of the event at the university and a brief but yet powerful message addressed to the PRI leadership and to the media of "dubious neutrality": they stated that they were students from the *Ibero* and that nobody trained them for anything. This video was spread on social networks and the hashtag #131alumnosdelaIbero (#131studentsfromIbero) soon became a world "trending topic".

Days before this event, Twitter was also mobilized, on May 1, 2012; the two most important TV networks announced that they would not broadcast the presidential debate. That same night, @lvloon posted the following tweet: "who is for the #AntiEPNdemonstration. I'm in". This tweet was sent to Julio Hernández López, a recognized journalist of the leftist newspaper *La Jornada*. Within four hours, the hashtag #AntiEPNdemonstration (#MarchaAntiEPN) was retwitted 85 times. However, the #AntiEPNdemonstration did not appear until May 11, 2012, the same day in which the candidate of the PRI attended the Iberoamericana University. (De Mauleón, 2012). The final hashtag that gave the movement its public name was created by @aochoad, a student from the University of Guadalajara, who shared with his 126 followers the following message: "Congratulations to the 131 students from the Ibero, I am number 132!" ("¡Felicidades a los 131 estudiantes de la Ibero, yo soy el número 132!"). Denise Dresser, a well-recognized columnist, spread the hashtag five hours after (De Mauleón, 2012). The rest is a well-known story.

Due to the video's popularity, students organized a public demonstration in front of Televisa's headquarters, the largest TV network in Mexico, in favor of transparency and neutrality of the media. This was based on the assumption that this network had a clear tendency in favor of Enrique Peña Nieto's electoral campaign. Protesters also asserted that most of the information online was unreliable due to the authority given to traditional media (TV and radio networks). This demonstration was directly linked to the previous hashtag #AntiEPNdemonstration.

From that moment on, the movement gained more and more followers and organized demonstrations in favor of the democratization of the media and against Enrique Peña Nieto. On May 30, 2012, the movement organized the "First General Assembly of University and Civil Society #YoSoy132" at the National University. At that time, the movement recognized the adherence of 54 public and private universities, 20 civil society organizations and the acknowledgement of 90 different local, national and international media representatives. The Assembly was attended by 6,000 people and was broadcasted on Internet (De Mauleón, 2012).

After seven hours of debate, the movement agreed on the following points: confirm the peaceful feature and the political neutrality of the movement; strong opposition to the imposition of any candidate elected through the mass media; rejection of the neoliberal economic model; protest against information bias by the traditional media and struggle for the exercise of an informed and free ballot, the statement that the movement was openly against Enrique Peña Nieto and his party and because his candidacy represented the old repressive and corrupt regime of the past and the factual powers that seek to impose him as the President of the Republic (González, 2012). It is noteworthy that the adherence of public university students to the movement meant the inclusion of social justice demands beyond the strictly political claims voiced by private university students (Olivier, G., 2015).

During the following weeks, the movement transcended social media and was already in the public debate. Its direct impact consisted in bringing *flavor* to a predictable and boring electoral campaign. However, its most important success was that *#YoSoy132* managed to organize a third presidential debate among all candidates and without the interference of the Federal Electoral Institute. Peña Nieto did not attend the debate. However, it proved the force of the movement and its capability for permeating and transcending social media. Another important success of the movement was that it managed to organize 3,000 electoral observers throughout the country during the electoral day.

Apart from this, it is important to stress the role that social media played in organizing the movement and the use its members and supporters made of Internet platforms. From the very outset, the movement disseminated its message through Facebook and Twitter accounts; they combined YouTube, Vimeo and other social media to upload videos, pictures and all sorts of communications. Twitter was mainly used as a quick and accurate source of information, whereas Facebook acquired a genuinely political status, as it was the platform for organizing the movement. Through social media they organized several demonstrations that were larger and larger in number. Other social movements such as the Movement for Peace with Justice and Dignity and labor unions joined the demonstrations.

The alliances and narrative structure adopted by #YoSoy132 weakened the movement, alienating private university members and radicalizing the repertoire of actions. Once the movement allied with unions and long-established movements and adopted assembly-like decision-making processes, private university members felt gradually alienated and after the electoral victory of Enrique Peña Nieto, most of them decided to leave the movement. Some of the remaining members advanced an anti-neoliberal agenda and therefore staged resistance and confrontational actions against the government leading to the clashes with police forces during the takeover ceremony of Enrique Peña Nieto.

Despite the lack of a well-defined leadership within the movement, some prominent members took advantage of their media exposure and successfully advanced their careers later on. Antonio Attolini, initially self-proclaimed spokesman of the movement, joined a TV program called "Sin Filtro" (broadcasted in Televisa-owned Foro TV channel), advised a senator and currently is coordinator for external affairs at the Mexican Social Security Institute. Carlos Brito, one of the moderators in the presidential debate organized by #YoSoy132, ran for the elections of his municipality and became independent (non-partisan) local mayor from December 2018. Ana Rolón, one of the producers of the movement's triggering video, currently owns a communications agency that supports civil society organizations. Even though the movement did not achieve its original goals and eventually dissolved, some members managed to boost their profiles in the political and civil society arenas.

#YoSoy132 sprang up in social media, and although it did not change the final results of the election and its influence and power was diminished after the election, it was unique and different from previous social movements in Mexico in several ways: it occurred few months before an electoral process that was supposed to be predictable: nobody doubted at the time that Enrique Peña Nieto was the virtual winner. It was initially a movement organized by students from private universities, a sector of the population traditionally not politicized. #YoSoy132 challenged the *status quo*; raised an important attention to the electoral process and managed to get the "democratization of the media" --*i.e.* regulate the TV concessions and create a more competitive market-- into the public agenda. #YoSoy132 also forced Televisa to change its electoral coverage and to provide members of the movement with a space to express their opinion. It also forced the PRI to change its speech and to incorporate the "democratization of the media" as part of its electoral proposals. And most importantly, #YoSoy132 proved that social media is a powerful tool in mobilizing people and influencing the decision-making process.

## The Wirikuta Defense Front

Although Mexico has gone from a rural to an industrialized country, the pressure for land and biodiversity conservation remains an important issue in national politics. This problem is especially important in indigenous areas where the economy is still based on self-consumption and where the larger share of the impoverished population live. According to INEGI in 2015, 7% of the Mexican population spoke a native language.

One of the most ancient and important indigenous groups are the *huicholes*, or *wixáritari*. According to INEGI, in 2015, there where 52,483 people that spoke the *huichol* language (INEGI, 2015). The *huicholes* have been involved in land conflicts since the 19th century; this in turn has made land defense an important part of the *huichol* culture and cosmogony. Another important feature of the *huicholes* is that, unlike other indigenous groups, they are not settled in a specific geographical region but rather spread in the states of Jalisco, Nayarit, Durango, Zacatecas and San Luis Potosí. Thus, in order to preserve their culture, between October and March, the *huicholes*, make an annual pilgrimage to *Wirikuta*, a region near Real de Catorce in the state of San Luis Potosí, where they look for the *hikuri* or *peyote*, a cactus that is a central element of the *huichol* religion. However, the *huicholes*' lack of land rights over this territory has provoked clashes with local farmers and shareholders of common land. After several years of requests to authorities, in 1998, the government of San Luis Potosí officially recognized *Wirikuta* and the Historical-Cultural Route of the Wixárika People as a natural protected area. In understanding the *Wirikuta* Defense Front, it is important to keep in mind these facts.

As stated, the *huicholes* have been struggling to defend their land since 19th century. This struggle had been a local issue that had no attention from the federal government and that was not a priority in the public agenda. This changed in 2009, when the Federal government granted twenty two concessions (initially valid until 2060) to First Majestic Silver Corp. and Revolution Resources Corp., two Canadian mining companies, to exploit silver, gold and other minerals in 6,326 hectares in the state of San Luis Potosí, out of which 70% are in the *Wirikuta* Area.

Mining in the San Luis Potosí area is not a new activity, however, no project has been as large as and as ambitious as the concessions to the Canadian mining companies. In 2010, a movement called The *Wirikuta* Defense Front (Frente en Defensa de Wirikuta. *Tamatsima Wa haa*), opposed to the mining exploitation on the area, was created. This movement was formed by indigenous communities and micro-entrepreneurs, mainly from the tourist sector and local non-governmental organizations.

The movement was organized from a twofold perspective: on the one hand, the environmental protection and, on the other, a cultural heritage protection. According to the Front, the exploitation plan developed by First Majestic Silver would take place through chemical processes, where highly polluting chemical would be used, affecting existing aquifers that supply communities near Real de Catorce. Thus, the mining of First Majestic Silver jeopardizes both the land rights of indigenous peoples recognized in various international conventions, as well as the Inter-American System of Human Rights, including the right to indigenous cultural identity and freedom of belief of the *wixárika* people. The project would also violate the right to water, health and environment of the people living near *Wirikuta*, and their right to sustainable development (frenteendefensadewirikuta.org).

The Front's advocacy strategy was based on nine demands: 1) *Wirikuta* should be declared as part of UNESCO Network of Natural Sacred Sites and protected in consequence; 2) Cancelation of any mining activity in *Wirikuta* and the prohibition of any administrative permission that granted this activity; 3) No new concessions granted in *Wirikuta*; 4) The *Wirikuta* reserve should be declared as a Protected Natural Area; 5) All the Real de Catorce Mountain Range should be declared a Cultural Landscape; 6) The pilgrimage path to *Wirikuta* should be protected under the UNESCO's World Cultural and Natural Heritage Convention; 7) Environmental rehabilitation should be started in the *Wirikuta* area; 8) Preventive actions should be ensured to protect the health of the inhabitants of *Wirikuta*; and 9) Allocation of federal and local resources aimed at improving the living conditions of the rural population of the region should be granted (Wirikuta's Defense Front, n.d.).

According to their website, the *wixaritari* authorities elected a local organization that has defended the agrarian rights of the indigenous population (Asociación Jaliscience de Apoyo a Grupos Indígenas, AJAGI) as their representatives in the defense of *Wirikuta*. This organization included other local organizations and together they formed the Front. The main objective of the Front was to get public attention in order to reach agreements with institutions, obtain legal advice and pressure the executive and the legislative to design and implement policies for the protection of *Wirikuta*.

The Front is composed of five Committees: political and legal; environmental; communitarian work; artistic promotion and communication (Álvarez, 2012). This institutional design is in line with the Front's advocacy strategy that has two main features. On the one hand, there is a legal component that seeks, in the short term, to settle the dispute before Federal Courts to cancel mining concessions granted to Canadian companies. In the long term, the legal component intends to protect *Wirikuta* at the international level by changing the very nature of the development plan designed by the Federal and local governments through declaring the area as a cultural and national heritage ensuring that no mining or any other industrial activity will be carried out in *Wirikuta*.

The other part of the strategy has to do with fundraising and keeping the issue on the public agenda. In achieving this, the Front designed an advocacy program based on the dissemination and socialization of the *huichol* culture, its symbols and rites combined with the support of strategic allies in the cultural, academic and political arenas. The strategy used social media as the main tool for raising public awareness and achieving the overall goals of the movement. From 2010 to 2014, the Front produced several short films and a full-length film with widespread international exposure through social media, which tells the history and traditions of the wixárica people: *Huicholes, los últimos guardianes del peyote* (Gasparello, 2016).

2011 and 2012 were two very important years for the Front, because it managed to gain public attention at the national level by organizing three major events. The first one was a demonstration in October 27, 2011 in Mexico City as part of the advocacy activities to raise awareness on the importance of protecting *Wirikuta*. The demonstration sought to emulate the *huichol* pilgrimage to *Wirikuta*, so it gathered actual priests and people who shared ritual objects and chants with attendants to the demonstration. The demonstration started in Reforma, the most important avenue in Mexico City, and ended at Los Pinos, the then presidential residence, where *huichol* authorities delivered their advocacy points to the President. The second important moment took place in the Senate, where the Front organized a two-day public debate aimed at creating a legal proposal in defense of *Wirikuta*. *Huichol* authorities, advocates, academics and representatives of the three main political parties in Mexico attended the event. The representative of these parties publicly agreed to present a bill in favor of *Wirikuta* to the Congress.

Finally, the Front organized a massive concert in *Foro Sol*, the most important concert venue in Mexico City, on May 26, 2012. The concert was intended to raise awareness on the defense of *Wirikuta* among young population and also to serve as a fundraising campaign. The *Wirikuta Fest*, as it was called, was a major event, since it brought together iconic Mexican rock groups and international personalities such as Caifanes, Café Tacvba, Ely Guerra, Enrique Bunbury, Calle 13 and Julieta Venegas. This event marked a turning point for the movement, since it turned the Wirikuta case from a local conflict to an increasingly political issue inserted in the public agenda.

The dispute was temporarily settled before Federal Courts in February 2012 and in September 2013, supporting the idea that the Front succeeded in suspending, at least temporarily, 78 concession project in the sacred area of Wirikuta, but the conflict persists with no visible end in sight. Even though the mining projects are stalled, the mining companies still conduct public relations activities, under a development mantra, in the region to improve their image vis-à-vis the local population and they support local organizations that put pressure on the Wixarika Regional Council to allow the operation of mining activities in Wirikuta. In recent months, the Wixarika-led organizations have asked for an audience with the left-wing president Andrés Manuel López Obrador to address the original demands of the Wixarika people for the preservation and protection of their natural and cultural heritage through the implementation of conservation projects that could provide jobs for the local impoverished inhabitants.

The Front implemented a strategy that proved to be very effective to insert an apparently local problem into the national agenda. This strategy included a constant support of well-recognized cultural, academic and activist personalities, a strategic use of social media, through websites, blogs, Twitter and Facebook accounts and a clear and direct message to local and national and policy-makers. This strategy was built upon *huichol* symbols, rites and icons that made supporters identify with the *huichol* culture.

The movement for *Wirikuta* has a different nature and was not created in social media but it was enhanced through it. This is an interesting example on how social media can turn a local problem into a national one and also because it shows how the use of Internet can benefit, support and advocate for a cause of an impoverished community that has no access to Internet, hence, that are not direct users of ITs.

These examples on the use of social media to organize political mobilization have common features that should be stressed. The first evident one is a strategic use of social media to disseminate a clear message in favor of their causes; to organize mobilization and to design its advocacy campaign. In the case of #YoSoy132, the movement intended to move electoral discussions out of the traditional channels and to insert media democratization in the electoral agenda. The *Wirikuta* movement intended to raise awareness about an environmental and cultural conservation holy site. Both movements started in a very local and specific community (students and indigenous); however, social media served as a useful and important tool for bringing these specific issues into the national political agenda. It seems that #YoSoy132 was a spontaneous movement with little time to plan and elaborate a long-term agenda. Maybe this explains why the movement lost its *momentum* and did not heavily impact the course of the presidential election. #YoSoy132 did not turn to be the "Mexican Spring" as some academics, columnists, experts and activists foresaw in May 2012. The *Wirikuta* Movement, in contrast, was organized around a clear and well-planned strategy. This also explains why they achieved specific goals: a political commitment from Congressmen of the three main parties to regulate mining exploitation in the San Luis Potosí region; a well-structured legal defense in favor of *Wirikuta*; and a settlement of the issue before Federal Courts that temporarily suspended the exploitation of the mines.

Finally, there is another common feature of both movements that unfortunately, has not gained enough attention in the academic discussion: the psychological aspect and how attractive it was to be part of any of these movements. Both the Movement in Defense of *Wirikuta* and #YoSoy132 generated a sense of belonging. They created a common powerful message: in a time were nothing surprises us anymore, there is still hope and collective action can (and to some extent, did) accomplish the impossible: it is worthy to fight for a cause that transcends social boundaries, personal beliefs or geographical regions. Belonging, engaging or simply sympathizing with these movements appeared to be seen as socially and politically correct. Attending demonstrations or cultural events related to these social movements was *cool* and *trendy*. Students were organizing a huge social party and everyone was invited. In words of a #YoSoy132 leader from the Instituto Tecnológico y de Estudios Superiores de Occidente (ITESO):

*#YoSoy132 has mobilized people very fast. Something happens and there is solidarity, the brotherhood of all, I think that this is the positive feature of the Movement. We had never met before, but a person comes and another joins, this simple fact, that were are here for a common cause, makes you fell that you have met your colleagues for a long time; and you trust them. So I think that this feeling has driven us. It's just like a cyclist: when you are down on the street by yourself and you feel weak, you feel that at any time a car is going to run over you... But when you join a ride with others the feeling is different. It's like "Oh! Cool! I am not alone, nothing bad could happen". Same thing happens here: if you are alone against the system, you*

*are in danger, but if you are part of a crowd, you are safe. It gives you strength and courage (Díaz, 2013).*

The Front in Defense of *Wirikuta* was organizing a different party: it created a place where you can feel protected, a place where globalization met sacred and ancient traditions, a place where you can go and be part of an amazing heritage. The coolness of the message was that anyone could defend *Wirikuta*, regardless of his race, social class, age or even nationality. As the Front's anthem states[1]:

*Wirikuta* is not for sale, *Wirikuta* is a place to defend
*Pachamama* warriors, love for my people listening to the desert, I found myself, ancestral forces guided my way.
It cannot be sold, it cannot be killed, they could have a lot of money, but here it is worthless [...]
In this embrace, we dance, we offer what we have from the heart [...]
We are calling neighborhoods, countryside and the city to join us,to join the awakening of my sacred heart homeland
*Wirikuta* is in danger!, general alert! The merchants of the earth want to plunder it [...]
This fight is for life, our natural temple together we will achieve to stop this destruction [...]
if you add your voice, this is stronger (*Wirikuta* no se vende, AHO Colectivo).[2]

This, of course, explains why these movements were so appealing to students, the young population and activists: simply, they were very *cool* and the song is catchy. This common feature also explains why these movements gained such an outstanding media coverage. The #WirikutaNoSeVende (*Wirikuta* is not for sale) and #YoSoy132 hashtags went from web-based viral trending topics to social trending topics. At the time, it was impossible to be uninformed about what was going on.

However, it is noteworthy that, at its zenith, the two movements shared the same fate: they welcomed other movements and social causes that, on the one hand, expanded its social base, but on the other, diluted the message and the cause for which they were advocating for in the first place. This seems an inevitable trade-off that any social movement should eventually face: the more supporters from different arenas, the more the messages that are diluted during the advocacy campaign. The question that should be further explored is to what extent do social media accelerate this phenomenon. It might be argued that in the absence of an institutional network that organizes mobilization, and without a clear plan to translate demands into specific public policy actions, web-based social movements are ephemeral. In the case of Mexico this argument holds: the #YoSoy132 movement diluted after elections; the message was not translated into a long-term public policy that changed the rules of media regulation in Mexico. The Front in Defense of *Wirikuta* is still active and it managed to stop mining exploitation, at least, for a while. Nonetheless, the fate of these movements were not as tragic as their MENA counterparts where the unleashed social fever led to political instability and to a resurgence of authoritarian practices to ensure social order.

Finally, we should keep in mind that these two movements emerged in a specific political context: an electoral year. This is important because during elections, citizens tend to be more politicized than in other cases, hence, they are more willing to engage in political mobilization. The relationship between the use of social networks and the degree of political participation should be analyzed in a less-politicized time period to start drawing some plausible and well-sustained conclusions in this arena.

## THE ROLE OF SOCIAL MEDIA IN THE CONSOLIDATION OF DEMOCRACY IN MEXICO

A slogan of a very famous radio station stated many years ago: "the music that is here to stay". In the same sense, social media is in Mexico to stay. As in other cases, empirical evidence to support the argument that the use of social media enhances political and social participation and mobilization is yet to be systematized. However, the undeniable fact is that social media now plays a crucial role in the way in which the State, the media and the traditional institutions of political representation relate with their public, their constituencies and their citizens.

After #YoSoy132 and the Movement in Defense of *Wirikuta*, things changed. Now there is no government institution, no political party, no media broadcaster, no public figure that does not have an active army of community managers dedicated to create the adequate message, that look for the right time, and that search for the key public to communicate what they want to on Twitter, Facebook, Google+, YouTube and so on. Only those who love political traditions could deny that the public agenda is now created, discussed and even implemented in social media. Political and social arrangements have changed since the last time PRI was in power: electoral competition is stronger at all levels and branches of government; the digital divide is narrowing; and people have a new and relatively attainable tool to share information, express their ideas and to organize and mobilize. These factors are becoming a more and more significant instrument for holding governments accountable and avoid the concentration of power, even though at the same time governments have the ability to censor and block content or disseminate fake news to further their own political agendas.

This is especially important in the current context. Andrés Manuel López Obrador came to power in 2018 with the promise of curbing corruption, bringing social justice and decreasing violence. The relationship between State and citizens is changing, as the president addresses the people directly and develops a direct channel of communication, which pushes state institutions into the background. Consequently, some political and social groups are feeling displaced and have shown discomfort and opposition against what they perceive as a polarizing narrative. Simultaneously, the gradual concentration of power in the president may lead to authoritarian practices that could threaten democracy, free speech and civil liberties. The challenge for new (and old) social movements is how capable they are to transform social mobilization into institutionalized practices that translate its social causes into specific, attainable and useful public actions.

It is yet to be seen if social media will consolidate as a sufficient condition to curb undemocratic practices, avoid the concentration of power and fight against the monopoly of information that the two main broadcasters have. However, it is possible to state that the use citizens give to social media has changed the way governments relate with their constituencies. Some would say that #YoSoy132 and the Front in Defense of *Wirikuta* were ephemeral. From a broader perspective, the mere fact that they changed the political agenda by exploiting social media was a big deal. Even if they disappeared as influential social movements, it really does not matter. It was worth it.

## REFERENCES

Álvarez, I. (2012). *El frente en defensa Wirikuta Tmatshima Wahaa: un acercamiento sistémico a los Nuevos Movimientos Sociales* [Wirikuta Tmatshima Wahaa's Defense Front: A Systemic Approach to New Social Movements] (Unpublished Graduate Dissertation). National Autonomous University of Mexico, Mexico City, Mexico.

Álvarez, I. (2019). Wirikuta en disputa. Reivindicación del patrimonio ejidal y minero en el Altiplano Potosino [Wirikuta in dispute. Claim of the ejido and mining heritage in the Potosian Plateau]. *Sociedad y Ambiente Society and Environment*, *7*(19), 31–54.

Bernard, E., Steen-Johnsen, K., & Wollebaek, D. (2012, January), *How do social media change the conditions for civic and political mobilization*? Paper presented at 3rd. International Conference on Democracy as Idea and Practice, University of Oslo.

Breuer, A. (2012). *The Role of Social Media in Mobilizing Political Protest. Evidence from the Tunisian Revolution*. Discussion Paper 10/2012, German Development Institute.

Coleman, S. (2001). The Transformation of Citizenship? In B. Axford & R. Huggings (Eds.), *New Media and Politics* (pp. 110–121). London: SAGE Publications. doi:10.4135/9781446218846.n5

De Mauleón, H. (2012, Sept. 9). De la red a las calles [From Social Media to the Streets]. *Nexos*.

Díaz Alba, C. (2013). Tres miradas desde el interior de #YoSoy132 [Three looks from inside #YoSoy132]. *Desacatos*, 43.

Gladwell, M. (2010, Oct. 4). Small Change: Why the Revolution Will not be Tweeted. *The New Yorker*.

González, J. L. (2012). #YoSoy132: Participación política 2.0 en México [#YoSoy132: Political participation 2.0 in Mexico]. *Diálogo Político*, *29*(3), 77–104.

González S., Borge-Holthoefer, J, Rivero, A., & Moreno, Y. (2011). The Dynamics of Protest Recruitment through an Online Network, *Scientific Reports*, *10*(1038), 1-7.

Lamberti, M. J. (2011). *"Abajo del amanecer" el corazón del universo en disputa. El caso de la instalación de empresas mineras en Wirikuta* ["Down the Dawn" the heart of the Universe in dispute. The case of the instalation of mining companies in Wirikuta] (PhD dissertation). El Colegio de México.

National Institute of Geography and Statistics. (2004). La población Indígena en México [Indigenous Population in Mexico]. Author.

National Institute of Geography and Statistics. (2005). *INEGI*. Aguascalientes, Mexico: Conteo de Poblacion y Vivienda.

National Institute of Geography and Statistics. (2010). *INEGI*. Aguascalientes, Mexico: National Census of Population and Living Conditions.

Norris, P. (2002). *Democratic Phoenix. Reinventing political activism*. Cambridge: Cambridge University Press. doi:10.1017/CBO9780511610073

Owen, D. (2017). The New Media's Role in Politics. In *The Age of Perplexity. Rethinking The World we knew*. Madrid: BBVA.

Saavedra, M. (2014). Sistema de protesta: política, medios y el #YoSoy 132 [Protest system: politics, media and the #YoSoy 132]. *Sociológica*, *29*(82), 83–123.

Stanley, J. W., & Weare, C. (2004). The Effects of Internet Use on Political Participation: Evidence from an Agency Online Discussion Forum. *Administration & Society*, *36*(5), 245–269. doi:10.1177/0095399704268503

Turnsek, M., & Jankowski, N. W. (2008, April). *Social Media and Politics: Theoretical and Methodological Considerations in Designing a Study of Political Engagement.* Paper presented at Politics Web 2: An International Conference, New Political Communication Unit, Royal Holloway, University of London.

Valenzuela, S. (2013). Unpacking the Use of Social Media for Protest Behavior: The Role of Information, Opinion Expression and Activism. *The American Behavioral Scientist*, *XX*(X), 1–23.

Vega, A., & Merino, J. (2011). *Ciudadanos.mx* [Citizens.mx]. Random House Mondadori.

Wirikuta's Defense Front. (n.d.). Retrieved from http://www.frenteendefensadewirikuta.org

World Bank. (2019). *Individuals using the Internet (% of the Population).* Retrieved from World Bank Open data: data.worldbank.org

## ENDNOTES

1 Aho Colectivo, an *ad hoc* group created by popular personalities in favor of the Movement, composed the Front's anthem as part of its fundraising campaign. The anthem was sold on iTunes.

2 Spanish lyrics: "*Wirikuta* no se vende, *Wirikuta* se defiende / *Pachamama warriors,* amor para mi gente / escuchando al desierto, me encontré a mí mismo fuerzas ancestrales que guiaron mi camino. No lo pueden comprar, no lo pueden matar, por mucho dinero que tengan aquí no vale nada [...] En este abrazo, vamos danzando, lo que tenemos de corazón lo ofrendamos [...] Llamando a los barrios, al campo y la ciudad, al corazón sagrado de mi patria despertar. ¡*Wirikuta* está en peligro, alerta general!, los mercaderes de la tierra la quieren saquear / Esta lucha es por la vida, nuestro templo natural. Unidos lograremos esta destrucción parar [...] si sumas tu voz esto cobra más fuerza".

# Chapter 4
# The First Amendment in the Digital Age:
## What Is Speech and How Should It Be Protected?

**Sam B. Edwards**
https://orcid.org/0000-0002-5911-0031
*Quinnipiac University, USA*

## ABSTRACT

*The United States is facing challenges in applying First Amendment principles from the 18th century to modern communications. This chapter examines cases where the government has intruded upon First Amendment rights. The first section examines when the government attempts to prevent protests by cutting internet access. This amounts to a digital gag and ear plugs for the protesters. Aside from cutting access to a single area, some governments have access to total internet "kill switches." This allows unsurpassed censorship of speech. Now there are technical means installed in most phones that could allow governments to disconnect internet access at the individual phone. In the area of social media, the courts are struggling to identify what constitutes speech. For example, when is a "like" or a "wink" speech? Equally important, when can a political figure censor speech through blocking users on social media? These cases represent warning signs that the United States, just like other countries, is struggling to adapt eighteenth century legal principles to modern communication.*

DOI: 10.4018/978-1-7998-3152-5.ch004

## INTRODUCTION

According to the International Telecommunications Union, 3.9 billion people are online. That represents the majority of the world's population. The ITU notes that growth in developing countries is reaching saturation while there is still "ample room for growth" in least developing nations. (International Telecommunications Union, 2018). The ability of so many people to communicate, share ideas, and organize is unprecedented. Along with this interconnectedness comes the power to challenge governments. Although the United States has long protected political speech and the right to organize protests, the digital age has provided a platform that was unfathomable when the founders created the United States. The government is struggling in applying the concepts of freedom of assembly and speech to situations made possible by digital communications. Ultimately these tensions end up in the legal system and courts are trying to define how these concepts apply considering all of the technical advancements in communication.

Advancements in technology have made the *speed* and *scale* of communication jump exponentially. Although technology has eroded privacy, it has also made it possible for people to share ideas and organize protests against the government at unprecedented speed and scale. To what extent are these fundamental rights scalable? Are they absolute or do they need to be modified given the scale and scope of communications?

When the founders of the United States incorporated the fundamental principles of freedom of speech and assembly into the US Constitution, they understood the importance of these rights in a democracy. Although these rights are essential to a democracy, they also pose threats to a government as they give power to the people to affect change. This tension between what is best for a democracy and what is best for a current regime underlies many efforts by governments to curtail these rights in order to maintain power. The Constitution protects the rights of people to organize and protest the government and share their ideas with others.

The United States Constitution provides "Congress shall make no law...abridging the freedom of speech, or of the press; or the right of the people peaceably to assemble and to petition the Government for a redress of grievances" (Amendment I).

### Why Protect Freedom of Speech?

There are many reasons the founders chose to protect freedom of speech and assembly. First, the "First Amendment was adopted to curtail the power of Congress to interfere with the individual's freedom to believe, to worship, and to express himself in accordance with the dictates of his own conscience" (*Wallace v. Jaffree*, 1985).

Second, the First Amendment protects "unpopular individuals from retaliation -- and their ideas from suppression -- at the hands of an intolerant society" (*McIntyre v. Ohio Elections Commission*, 1995).

Third, the court ruled that, "each person should decide for him or herself the ideas and beliefs deserving of expression, consideration, and adherence...government action that stifles speech on account of its message or that requires the utterance of a particular message favored by the government, contravenes this essential right (*Turner Broadcasting System v. Federal Communications Commission, 1994*).

Thus, it is important that people be able to express themselves without fear of reprisal from the government, because a democracy is most successful when people are freely able to exchange ideas and form their own opinions. These protections are important to a democracy because they allow for a free exchange of ideas, and from this "marketplace of ideas," a society can select its path. "The First Amendment reflects a profound national commitment to the principle that debate on public issues should be uninhibited, robust, and wide-open, because speech concerning public affairs is more than self-expression; it is the essence of self-government" (*Snyder v. Phelps*, 2011).

Despite the importance of free speech and assembly in the United States, digital communications are complicating these rights as applied. What does it mean to have free speech in the age of instant and constant communications? What constitutes speech when we have so many ways of communicating? These are the issues underlying freedom of speech and assembly in the digital age.

This chapter will examine how the US Constitution applies in the context of free speech and assembly in this new digital age.

Section I explores the right to assemble and freely exchange information and intentional wireless interruptions by the government. Political protests are now organized and carried out through social media rather than through face to face meetings. Again, the current legal system is struggling to apply the First Amendment to this new digital world.

Section II examines to what extent the Constitution protects political speech when that speech is digital in nature. Speech in the digital age is quite different from when the First Amendment was debated and the current legal system is facing challenges adapting this right to the digital world. The ability of political figures to communicate with constituents is essential to democracy but presents novel challenges on the Internet. On the one hand, private platforms such as Twitter allow politicians to circumvent traditional media and take their messages directly and immediately to the public. These same platforms also allow the government officials unprecedented technical means to block and silence dissenters.

The chapter concludes with predictions where this debate over freedom of speech and assembly is likely to go in the future.

## SECTION I: WIRELESS INTERRUPTIONS AND THE FIRST AMENDMENT

"The first thing we do, let's kill all the lawyers" is the first act a tyrannical ruler seeks to do because lawyers are the defenders of democracy. (Shakespeare, 1598, 4.2, 73). While this advice is still valuable for current tyrannical rulers, a more modern version might be "The first thing we do, let's shut off their internet access." By removing the ability of those who oppose a government to communicate, assemble, and protest the tyrannical ruler can accomplish what Shakespeare's character wanted to do in the 1500s much more expeditiously and without the bloodshed.

When the founders created the Constitution of the United States, they understood the value of freedom of speech to a democracy. Without the ability of people to organize and freely exchange ideas, a true democracy could not flourish. For these reasons the First Amendment protects those most important of rights to a democracy, freedom of speech and assembly. In fact, freedom of speech was so central to the founders' idea of a democratic government that they did not feel the need to include that right in the Constitution,

*I go further, and affirm that bills of rights, in the sense and in the extent in which they are contended for, are not only unnecessary in the proposed constitution, but would even be dangerous. They would contain various exceptions to powers which are not granted; and on this very account, would afford a colourable pretext to claim more than were granted. For why declare that things shall not be done which there is no power to do? Why for instance, should it be said, that the liberty of the press shall not be restrained, when no power is given by which restrictions may be imposed? (Hamilton, 1778, 575-81).*

Here Hamilton was expressing his view that the Constitution did not need the rights later included in the Bill of Rights because the Constitution, by its very nature, did not give the government the power to curtail the fundamental rights of freedom of assembly and speech. A government that does not allow free exchange of ideas and protests is not a democracy envisioned by these founders.

However, today through the advancement of communication technology people possess the ability to organize protests and share information at a speed and on a scale that gives citizens much more power than in the days before the Internet. Many governments have reacted to this power in the hands of citizens by attempting to prevent them from communicating. For example, the Turkish government blocked the social media service Twitter in order to prevent people from sharing information about a government corruption scandal (Arsu, 2014). China's "Great Firewall" is also a well-known example of a government trying to prevent citizens from sharing and accessing information. Researchers examined over 600 recent shutdowns of communication networks and concluded that these shutdowns most often occur in authoritarian regimes (Howard, Agarwal & Hussainet, 2011). These attempts to thwart citizens' access to and ability to share information run counter to the tenets of the freedom of speech and assembly.

Although these types of shutdowns occur most often in under authoritarian governments, the United States is not immune from wishes of government officials seeking to prevent people from sharing information and organizing protests. This section will examine where the government has sought to prevent people from accessing information and organizing protests.

## The BART Wireless Interruption

In the early morning hours of New Year's Day in 2009 a white San Francisco Bay Area Rapid Transit (BART) police officer shot and killed Oscar Grant III, an unarmed black man, at a transit station (Bulwa, Buchanan & Yi, 2009). The shooting and subsequent case fueled racial tensions and sparked protests (Leonard, 2010). People began to protest against BART and a group called "No Justice No Bart" began to organize and their demands included:

*total transparency in the investigation of - and accountability for - police misconduct. We want all available information released to the public when there are questions about whether the police acted appropriately, and we don't want any lies, half truths, creative ommissions (sic), or misdirections (No Justice No BART, 2014).*

This group organized protests against BART because of the shooting of Mr. Grant. As the protests gained momentum the protests escalated to the point where they planned to march on BART stations. In response, the government officials at BART shut down the mobile phone networks and Wi-Fi access points at the stations to prevent flash mobs from organizing (Policinski, 2011). BART officials claimed that the service interruption was necessary to protect the public during the busy rush hour time (Policinski, 2011). Internal messages by BART officials including Mr. Johnson the BART spokesperson indicate a different motive,

*A whole heck of a lot their ability to carry out this exercise is predicated on being able to communicate with each other. Can't we just shut off wireless mobile phone and Wifi communication in the downtown stations? It's not like it's a constitutional right for BART to provide mobile phone and Wifi service. Additionally, the wireless phone companies and Wifi Rail rent space from us. We have radios — seems to me we have the upper hand communication wise (Elinson, 2011).*

BART shut down the wireless networks and protest never happened (Elinson, 2011). Was Johnson correct that BART order the shutdown of mobile and Wi-Fi communication networks to stop a protest? What would the founders of the Constitution think of this use of government power to thwart citizens from organizing, sharing information, and protesting their government? Just like banning people from meeting to organize protests, BART violated the freedom of speech when it used this digital means of prohibiting protests.

BART officials claimed that their act of shutting down the communication systems was because of a threat to public safety (Policinski, 2011). This raises the question was this a time, place, and manner restriction independent of the content of the speech? Also, was this an appropriate prior restraint on speech? Put another way, did BART violate the free speech and assembly rights of the protesters or did its actions fit within exceptions to these rights? Unfortunately, this matter was never litigated so we do not know how a court would respond to these questions. This does set a dangerous precedent where the government can feel free to silence speech and the means to assemble and petition the government.

## Content Neutral Time, Place, and Manner Restrictions

The court has recognized that speech has limits (*Roth v. United States*, 1957). One such limit is on the time, place, and manner of speech.

*Expression, whether oral or written or symbolized by conduct, is subject to reasonable time, place, or manner restrictions. We have often noted that restrictions of this kind are valid provided that they are justified without reference to the content of the regulated speech, that they are narrowly tailored to serve a significant governmental interest, and that they leave open ample alternative channels for communication of the information (Clark v. Community for Creative Non-Violence, 1984.)*

Thus, BART's actions could be constitutional only if (1) its restrictions were unrelated to the content of the speech and based *entirely* on the need for public safety; (2) the restrictions were narrowly tailored to serve the government interest; and (3) BART left open ample alternative channels for communication. BART's actions fail all three parts of this test.

First, BART's argument that its actions were content neutral is dubious at best. Here the fact that BART responding to protests against it at least raises the concern that BART was trying to stop the protests. Moreover, the comments of the BART spokesperson indicate that they were looking to stop the protesters from organizing and they did not see any constitutional right to communicate. The court will look beyond the text of a restriction to find the actual intent. It is likely that the court would find that BART's motivation for the wireless interruption went beyond safety and was designed to prevent the protests.

Second, even if this was content neutral it still fails the test for constitutionality because the restriction was not narrowly tailored. By shutting off all mobile networks and Wi-Fi access, it prevented *all* speech, not just the speech of the allegedly dangerous protesters (Fleming, 2012). "BART's broad shut-down of its wireless network far exceeded the scope necessary to achieve its safety purpose. Not only did BART's actions reach its goal of safety, it also prevented all speech, including everyday speech - a father's call home to his family, a teenager's text message to friends, and 911 emergency calls" (Fleming, 2012, p. 645). In fact, by cutting off all communications BART actually made the situation *more* dangerous by eliminating the main way that people report emergencies, by calling 911 on their mobile phones. "The Commission has clearly made access to wireless emergency services a priority by ensuring that CMRS users may easily dial 911 and by working to ensure that emergency services can be reached through a wide variety of other means" (Siy, Feld, Bankston, Llansó & Tien, 2012, p. 4). Thus, not only did the restraint on speech eliminate virtually all speech, it actually made the situation more dangerous.

Third, BART did not leave open alternate means of communication for the protesters. "Other means of communication to achieve the purpose of organizing would have been completely inadequate because NJNB's political speech depended on its ability to quickly communicate with individuals while being in the underground tunnels" (Fleming, 2012, p. 647). Without the ability to communicate in the BART tunnels, it was not possible for the protesters to engage in their political speech. There simply was no alternative method available to them.

Thus, BART's actions fail all three parts of the *Roth* test and a court should find these actions unconstitutional. This is the type of restraint on speech that the Constitution was designed to prevent. A group of people, upset at the government, tried to share information and organize a protest. BART, under a claim of protection of the public, silenced the protesters' voices.

## Prior Restraint

Even if this action somehow survived *Roth* analysis, it still is unconstitutional because it was a prior restraint on speech. Here BART officials sought to prevent protesters from organizing and sharing information *before* they started to engage in that speech. This preemptive attempt to stop the protesters' speech is an example of prior restraint. BART did not seek a judicial order, and did not even see this as a constitutional matter according to a BART spokesperson, "It's not like it's a constitutional right for BART to provide mobile phone and Wi-Fi service" (Elinson, 2011). This amounts to a prior restraint on speech.

The court has long recognized that prior restraints on speech are "the most serious and the least tolerable infringement on First Amendment rights" (*Nebraska Press Association v. Stuart*, 1976, p. 559). This is because "prior restraint, by contrast and by definition, has an immediate and irreversible sanction. If it can be said that a threat of criminal or civil sanctions after publication "chills" speech, prior restraint "freezes" it at least for the time" (N*ebraska Press Association v. Stuart*, 1976, p. 559). In fact, the court concluded in an earlier case, "any system of prior restraints of expression comes to this Court bearing a heavy presumption against its constitutional validity.... We have tolerated such a system only where it operated under judicial superintendence and assured an almost immediate judicial determination of the validity of the restraint (*Bantam Books. v. Sullivan*, 1963, p. 70, citations omitted).

The purpose of this presumption against prior restraint and the requirement for judicial review is rooted in the democratic concept of the marketplace of ideas.

*If the government stops speech before it is spoken, the ability to seek the truth is significantly impaired. Due to this realization, prior restraints represent the antithesis of the marketplace of ideas because it prevents communication from occurring at all, thus thwarting the uninhibited public discourse of an issue (Fleming, 2012, p. 640).*

Here BART officials used technological means to prevent protesters from organizing, sharing ideas, and protesting BART's actions. BART took this extreme step prior to the protests explicitly to prevent the protests. In so doing, BART was able to prevent the protesters from getting their message heard by others. This attempt to keep a political voice silent through technology is exactly the type of action the First Amendment was designed to protect against.

Because of the strong presumption against prior restraint, the court requires the government to seek judicial review of their actions *before* taking the extreme action of silencing speech. The Supreme Court in *Freedman v. Maryland* (1965) set out the procedural requirements the government must follow before engaging in prior restraint:

1. The burden of proving that the film is unprotected expression must rest on the censor...
2. Any restraint imposed in advance of a final judicial determination on the merits must similarly be limited to preservation of the status quo for the shortest fixed period compatible with sound judicial resolution...
3. The procedure must also assure a prompt final judicial decision, to minimize the deterrent effect of an interim and possibly erroneous denial...(p. 58).

BART officials knew of the protests ahead of time since the protests had been building since the shooting. There was plenty of opportunity for BART to seek a judicial order but not only did it fail to do so, it clearly did not see it as a necessity. Even if the situation was so dire that a prior judicial application was not possible, the action fails parts 2 and 3 of the *Freedman* test since the restraint was overbroad and BART never submitted it for final judicial review. Likely, BART did not see that its actions were an infringement upon freedom of speech and assembly. What BART did amounts to putting gags over the mouths and ears of protesters to prevent them from communicating. BART just did this virtually through technology. Is a virtual gag any more permissible than a physical one? Herein lies the crux of the problem—should a government be able to, without judicial review, remove the technical means for people to communicate and organize without conflicting with the Constitution? Just like the government cannot stop a printing press because it disagrees with the pamphlets being printed, BART cannot shut down digital communications to prevent protests. Although a court never was called to pass judgment on BART's actions, legal scholars have concluded that,

*Ultimately, the First Amendment prohibits public agencies from interrupting wireless service in all but the most extreme circumstances. Shutting down wireless communications networks will invariably suppress completely lawful, constitutionally protected speech on a massive scale; it is incredibly unlikely that the threat addressed by a communications network shutdown will ever be justifiably proportionate (Siy et al., 2012, p. 9).*

In response to the criticism of BART's handling of the protests, it developed a new policy where it tries to balance freedom of speech and assembly with safety and security interests. Specifically, it permits BART to temporarily interrupt cellular service when

*there is strong evidence of imminent unlawful activity that threatens the safety of District passengers, employees and other members of the public, the destruction of District property, or the substantial disruption of public transit services; that the interruption will substantially reduce the likelihood of such unlawful activity; that such interruption is essential to protect the safety of District passengers, employees and other members of the public, to protect District property or to avoid substantial disruption of public transit services; and that such interruption is narrowly tailored to those areas and time periods necessary to protect against the unlawful activity (Bay Area Rapid Transit, 2011).*

This policy addressed some of the concerns raised during the incident, though scholars and civil rights advocates still argued that it leaves open the possibility of suppression of constitutionally protected rights.

The policy was eventually superseded by California state law. California Penal Code § 11471 Interruption of communication service; prohibited and allowable circumstances prohibits interruption of internet and cellular services absent a court order or "extreme emergency" such as severe bodily injury or harm. This law would have prohibited the type of interruption that BART used to prevent the protests. The fact that the legislative system eventually was able to address this situation demonstrates that it is more important than ever that the public voice their concerns over these intrusions in order to get protections through the legislature. In this case it only took 6 years for the law to catch up to the technology, a significant improvement over the Katz and Olmsted cases discussed above.

## Internet Kill Switches

Aside from the ability to effectively turn off access points to the Internet, some governments have engineered entire "kill switches" to turn off all access to the Internet. In reality there is no actual switch; "governmental officials would need to contact Internet Service Providers (ISPs), who would then authorize their engineers to change the configuration of the Internet traffic flow to stop service to the populace." (Medows, 2012, p. 59-60). Egypt in 2011 employed just this tactic in response to protests (British Broadcasting Corporation, 2011, January 28). Egypt is not alone in this the approach to eliminating access to the Internet. Nepal, Burma, and China have used this drastic approach as well. (Medows, 202, p. 64).

The United States Senate debated the "Protecting Cyberspace as a National Asset Act of 2010" that would have authorized "…the President to issue a declaration of a national cyber emergency to covered critical infrastructure…" (Protecting Cyberspace as a National Asset Act of 2010, S.3480). Ultimately the bill did not pass, though it does highlight the political pressure to give the government the power to control access in times when it senses an "emergency".

Debate continues at the national level about the necessity for the ability to restrict or eliminate access to the Internet in certain situations. For example, in the 2016 presidential election "witnessed U.S. Republican presidential candidates' debate closing portions of the Internet to counter ISIS" (Smith, 2016).

Although the internet kill switch bill did not pass the Senate, the government has instituted local internet shutdowns. Smith (2016) discusses a case where the FBI, without a warrant, shut down internet access of someone it suspected of running an illegal gambling ring. Under the ruse of repairing the internet, agents entered into the defendant's hotel room and were able to collect evidence. The government argued that "DSL is not an essential service" (*United States v. Wei Seng Phua*, 2015). The district court did not have to address the issue of the internet interruption since the court found "under the totality of the circumstances here, the Government has not met its burden of proving by a preponderance of the evidence that Phua's consent was voluntary." The court did note, "policy concerns also weigh against allowing the government to use a ruse of this type. Cable, telephone, and internet services are widely used in America. Most reasonable people would invite a third party repair person into their home if they were led to believe it was necessary to fix a problem with those services." Thus, the government is not above removing internet access in order to conduct surveillance that would be otherwise impermissible.

## Cell Phone "Kill Switches"

There are several ways to interrupt people's access to the internet and communications. As seen in the BART case the government can shut down cell towers. In the internet kill switch cases the government can order ISPs to shut down access. The third method is to turn off access on individual devices.

In response to theft of expensive cell phones, various legislatures have attempted to address this problem by requiring manufacturers to install "kill switches" on their devices. Minnesota was the first state with such a law. (Schmitz, 2016). California followed Minnesota in 2014 and now requires that manufacturers "shall include a technological solution at the time of sale, to be provided by the manufacturer or operating system provider, that, once initiated and successfully communicated to the smartphone, can render the essential features of the smartphone inoperable to an unauthorized user" (California Business & Professions Code § 22761). Because of the size of the California cell phone market, this essentially becomes a national standard, not unlike auto emission standards. As such, most every cell phone sold in the US must have this capability. Civil rights scholars raised the possibility of government abuse at the time the law was passed (Schmitz, 2016). These concerns were addressed in the subsequent law restricting the government's ability to interrupt communications, as discussed above.

Although California restricts the ability of the government to interrupt internet access, see above, other states may not have such restrictions in place. The ability to cut off internet access as a capability within most every phone without adequate legal safeguards opens the possibility of abuse. This is an example of how a law designed to protect consumers can end up giving governments the ability to suppress speech and assembly.

Thus, in the United States the government is struggling to find the bounds of protection for speech and assembly in the age of instant communications. Freedom of speech was much simpler when speech was through word of mouth and printed materials. The scope and speed at which information was shared and opposition organized pales in comparison to current times when a single video of a police shooting can go viral and create the impetus for a protest in a matter of hours. The principles of freedom of speech and assembly apply equally now despite this exponential change in the scope and speed of communications. Just because the government possesses the technical means to prevent speech does not mean that it should be able to exercise that power. The Constitution prevents just that type of infringement. Democracy is stronger for the ability to share information and this should be zealously defended.

## SECTION II: FREEDOM OF SPEECH AND SOCIAL MEDIA

Not only have technical advances in communication created legal problems, such as when a government seeks to prevent protests by shutting down communications, it also has created many more ways for people to communicate. This technical expansion of communication means that "speech" is no longer limited to only a few modes. Although the rights to political speech to protest are at the heart of the First Amendment, what constitutes speech in the digital age? In 1778, there were only a limited set of ways to communicate and those only allowed communication between small groups of people. How should free speech be interpreted now that instant communications with far more people is possible? Does the right to free speech scale with technology? In short, what is "speech" as envisioned by the drafters of the First Amendment?

This section will examine a case of political speech involving the "like" button on Facebook and will show how courts are wrestling with how the First Amendment relates to modern modes of communication. Again, the advances in communication technology have created new ways for people to communicate and the current legal system is struggling to adapt old concepts to these new advances.

This dispute started in 2009 when six employees in the Hampton Sheriff's Office supported Jim Adams, one of the Sheriff's opponents in the upcoming election. Support for Adams included "liking" Adams' Facebook page. B. J. Roberts won reelection and subsequently did not decide to retain any of the six employees (*Bobby Bland, Daniel Ray Carter, JR., David W. Dixon, Robert W. McCoy, John C. Sandhofer, and Debra H. Woodard v. B.J. Roberts*, 2012, p. 601).

The six employees brought suit against the Sheriff, the defendant, claiming that they were fired because of their political speech and the defendant filed a Motion for Summary Judgment. At issue in the motion was the Sheriff's argument that there was no "speech" and therefore no constitutional protection for the actions of the employees (*Bland et al. v.* Roberts, 2012, p. 602).

## Liking Is Speech

The essence of Roberts' argument is that clicking a "thumbs up" icon on Facebook is not speech, not even symbolic speech. As social media platforms have exploded over the past several years, so have the ways people communicate online. Trends in communication indicate that digital citizens value frequent, instant, and short communications. These trends towards speedy communication mean that instantly people can convey their feelings to a wide audience. With this speed has come a preference for brevity. Twitter is a good example. All communications are limited to 140 characters (Twitter, 2014). 140 characters is short but it is still undeniably speech.

Facebook has taken things a step further. Rather than writing something, one can express support or interest by clicking on a thumbs up sign. This action of "liking" is at the heart of the dispute in *Bland v. Roberts*. At what point does speech become so brief that is ceases to be speech?

The district court answered that question, "It is the Court's conclusion that merely "liking" a Facebook page is insufficient speech to merit constitutional protection" (*Bland et al. v. Roberts*, 2012, p. 604). The court distinguished other Facebook cases because in those there was always a comment along with the act of liking.

The court went on to explain its ruling,

*Simply liking a Facebook page is insufficient. It is not the kind of substantive statement that has previously warranted constitutional protection. The Court will not attempt to infer the actual content of Carter's posts from one click of a button on Adams' Facebook page (Bland et al. v. Roberts, 2012, p. 604).*

Thus, the court was not persuaded that "liking" on Facebook is a type of speech. It is particularly telling that court would not infer the content of liking a political candidate's Facebook page. Although this was a case of first impression for a US court, the court's conclusion that a "like" is not speech is troubling. Any regular user of Facebook would have no trouble inferring the meaning of liking someone's page. Why the court reached the opposite conclusion indicates that courts do not have a fundamental understanding of how people communicate online.

The district court was wrong in its ruling and ultimately the appeals court overturned the decision (*Bobby Bland, Daniel Ray Carter, JR., David W. Dixon, Robert W. McCoy, John C. Sandhofer, and Debra H. Woodard v. B.J. Roberts*, 2014). The Court of Appeals for the Eleventh Circuit closely examined what it means to "like" a page on Facebook by referring to Facebook's own online material (*Bland v. Roberts*, 2014, p. 42-4).

First, just because speech is online does not mean that it is somehow different from a constitutional standpoint. Legal scholars note that, *"speech" that is not written or spoken remains within the purview of the First Amendment. This remains true in the twenty-first century, as courts have found various new technologies, media, and modes of communication to be deserving of constitutional protection. For example, video games, posting hyperlinks to the Internet, sharing photographs on Facebook, and uploading videos to YouTube have been held to be "speech" protected by the First Amendment (Sklan, 2013, p. 383, citations omitted).*

Thus, the courts have already recognized that modern forms of communication are speech and therefore are subject to the First Amendment. This indicates that the judicial system is well equipped to apply constitutional provisions to modern forms of communication. Despite this body of case law, the district court in *Bland v. Roberts* was unable to conclude that liking a Facebook page was speech.

## Symbolic Speech Can Be Protected

The district court should have found that a Facebook "like" itself was a form of speech. Even if it could not come to this conclusion, the court should have found that the action of clicking a thumbs up was at least symbolic speech. Just like raising one's middle finger to express displeasure with the draft, speech can be symbolic and still protected by the Constitution (*Cohen v. California*, 1971).

Here clicking a picture of thumb indicating a "like" for a political candidate is as much political speech as was the middle finger salute in *Cohen*. If flag burning, political signs, and sit-ins are all symbolic speech more common in previous generations, then a Facebook "like" of a political candidate is the modern equivalent. Facebook makes the point clear in its *amicus curiae* brief in *Bland v. Roberts (2012),*

*If Carter had stood on a street corner and announced, "I like Jim Adams for Hampton Sheriff," there would be no dispute that his statement was constitutionally protected speech. Carter made that very statement; the fact that he did it online, with a click of a computer's mouse, does not deprive Carter's speech of constitutional protection (Facebook, 2013, p. 3).*

Fortunately, for users of Facebook and the Plaintiffs in the case, the Court of Appeals appreciated that this is the modern equivalent of a political campaign sign in one's yard and found that, "Aside from the fact that liking the Campaign Page constituted pure speech, it also was symbolic expression. The distribution of the universally understood "thumbs up" symbol in association with Adams's campaign page...conveyed that Carter supported Adams's candidacy" (*Bland v. Roberts*, 2013, p. 386).

Thus, even if the court found that a "like" was not speech it still should have protected it as symbolic speech. The district court was unwilling to entertain this important point, yet any user of Facebook would readily agree that the thumbs up has, at least, symbolic meaning.

## Political Speech Must Be Protected

Finally, and most importantly, the speech at issue in *Bland v. Roberts* is that most important speech, political speech. The Supreme Court has made it abundantly clear that political speech is where constitutional protections of the First Amendment "are at their zenith" (*Meyer v. Grant*, 1988, p. 437). Here, just as in *Meyer v. Grant*, the plaintiffs were engaged in political speech through their support of a candidate for office. A free exchange of ideas, even though symbolic speech, is at the core of the First Amendment. Without protection for political speech, there can be no marketplace of ideas. Here, not only was the speech political, it resulted in retaliation from an elected official because it went against him. Just like BART officials disagreed with the protesters, Sheriff Roberts retaliated against his employees because of their political support for his opponent. In both of these cases, the government officials in question did not see any problem with their actions. Moreover, in the *Bland v. Roberts* case the district court agreed and failed to protect this most fundamental of free speech rights.

The court should have been more vigilant in protecting political speech because this is the most important speech to democracy and the type most likely to be restricted by authoritarian rulers. These are the reasons that the drafters of the First Amendment included these freedoms and the reasons that the Supreme Court has singled out political speech as most deserving of protection.

## Does Freedom of Speech Prohibit "Blocking" by Government Officials on Social Media?

Aside from identifying speech on social media platforms, what happens when government officials try to suppress of block speech in those forums? The courts are "all over the map" when it comes to addressing this question (Hidy, 2019). This section will examine the current state of the law on this area of political speech on social media platforms.

President Trump makes frequent use of the Twitter to communicate directly with the public. He can bypass traditional media outlets and take his message, 140 characters at a time, directly to the public. This allows him to quickly and without much input from his staff get his immediate thoughts out to the public. This type of real time access to the current thoughts of an elected official is novel and presents many interesting issues. Here however, the constitutional concerns arise when an elected official chooses to block a member of the public from following them. Twitter's help entry summarizes the effects of blocking someone,

Blocked accounts cannot:

- Follow you
- View your Tweets when logged in on Twitter (unless they report you, and your Tweets mention them)
- Find your Tweets in search when logged in on Twitter
- Send Direct Messages to you
- View your following or followers lists, likes or lists when logged in on Twitter
- View a Moment you've created when logged in on Twitter
- Add your Twitter account to their lists
- Tag you in a photo

In essence, the block function allows a public official to silence someone in the social media marketplace of ideas. Moreover, blocking someone amounts to viewpoint discrimination since the official will target speakers with dissenting views. The courts have wrestled with applying First Amendment doctrine to social media platforms.

President Trump on his official Twitter account, @realDonaldTrump, blocked certain people and they filed suit. The appellate court found,

*the President blocked each of the Individual Plaintiffs (but not the Knight First Amendment Institute) from the Account. The government concedes that each of them was blocked after posting replies in which they criticized the President or his policies and that they were blocked as a result of their criticism. The government also concedes that because they were blocked they are unable to view the President's tweets, to directly reply to these tweets, or to use the @realDonaldTrump webpage to view the comment threads associated with the President's tweets (Knight First Amendment Institute at Columbia University v. Trump, 2019).*

The court went on to find that the President acted in a public capacity and that the account constituted a public forum. When the President blocked the users he disagreed with, he engaged in viewpoint discrimination within that public forum in violation of their free speech rights. The President argued that by blocking certain people he was refusing to listen to them. The court agreed that there is no constitutional right to "require the President to listen to their speech." However, the court noted that by blocking a user you prevent them from speaking to any of the other participants in that discussion.

*Knight* is the first case at the appellate level on this issue. Two scholars recently noted, "Officials are cautioned to proceed with care in this new virtual forum. When deciding whether to block a user, these officials ought to give full thought to whether such a decision is based on constitutionally impermissible considerations, such as the user's expressed political views" (Ostrove & Leber, 2019).

While the court in *Knight* found in favor of the plaintiff, other courts have ruled that government attempts to silence or curtail comments on social media platforms did not violate the Constitution. In *Davidson v. Plowman* (2018) the 4th Circuit upheld a decision that permitted censoring of "off topic" comments on the government Facebook page of a commonwealth attorney.

Other lower courts in different jurisdictions have also ruled against plaintiffs who allege constitutional violations by government officials when they have been blocked or removed from social media discussions (Williams, E. 2019).

Courts have just started to address freedom of speech claims related to blocked social media accounts. Until there are more appellate rulings, this area of law will remain in flux. Courts are the first forums where these issues will be raised. Legislatures have the power to establish firm rules that expand protections for speech if the so desire. Rather than leave these decisions up to the patchwork approach through the judicial branch, the legislative branches should step in. Uncertainty can chill speech and create problems for officials. Just what is permissible is still unclear.

## FUTURE RESEARCH DIRECTIONS

There will always be a tension between technology and constitutional law. The common law is well equipped to address the unforeseen changes due to technology, but it does not mean that it will always get it right on the first pass. As technology brings new ways to communicate, organize, and protest the government, legal scholars must be on guard to seek to apply constitutional doctrine to protect that speech.

### Wireless Service Interruptions

Specifically, the courts and legislatures need to address if and when a government can shut down mobile phone and internet communications as BART did. Since that matter never went to court, the issue will happen again. It is important to seek a judicial determination so that this matter will be resolved. The subsequent legislation in California is a good model to consider in other jurisdictions.

The FCC recently solicited public comment on rules regarding wireless service interruptions by governments. It is important to follow these developments and provide comments as the government struggles to find a balance between public safety and freedom of speech. State governments are also moving towards creating legal standards for wireless service interruptions. It is important to follow these developments and respond accordingly.

A ballot measure in states with that form of democracy is another option for advancing this issue. California seems the likely first step given the uproar the BART case generated.

## Freedom of Speech in the Digital Age

Digital natives will increasingly come at odds with existing legal rules regarding how they communicate. Unfortunately, most of the people in decision making positions in the judicial, executive, and legislative branches are not digital natives. They, like the district court in *Bland v. Roberts*, will not always understand the digital generation. There will be other cases similar to *Bland* where something abundantly clear to digital natives, is incomprehensible to the current decision makers.

There will also be tension with existing transparency laws such as the Sunshine Laws (open meeting laws) that will be pushed to apply to modern communications. Is a Facebook discussion on a matter between elected officials a "meeting" under the terms of the Sunshine laws? Digital natives communicate this way. Is this a meeting?

There is also a current debate over net neutrality. Is there a right to the internet? Is there a right to an internet that does not distinguish content? Net neutrality is bound to end up the courts. This will be another case where the courts will try to apply existing law to the new communication's landscape.

Where will courts and hopefully legislatures draw the lines around suppression of dissent in social media forums? When is an account a public forum and when is one private? Since these accounts are on private platforms this distinction can be murky. When is a block a violation of speech and when is it an appropriate way to deal with a heckler? These questions will need to be answered definitively so that officials and the public can navigate these issues.

## Privacy

The same technology that makes it possible to share information also makes it possible for the government to spy on its own citizens. As the recent NSA scandal has shown, the government not only possesses the technical capability but it has the political will to collect as much information on its own citizens as it can. If the CIA can spy on the Senate Intelligence Committee (Miller, O'Keefe and Goldman, 2014) then who watches the watchers?

Courts will be called upon to rule on the extent to which the government may use technology to conduct surveillance on people both in the United States and abroad. Just because it can does not mean that is should be legal. This is a fertile area for additional research as society tries to balance the need for privacy with the fact that technology has famously made the end of privacy.

Also, privacy will need to be applied to privacy from non-governmental actors as well. As technologies such as Google Glass (Google, 2014) become more ubiquitous society will have to define rules to protect privacy while recognizing the value of this technology.

## CONCLUSION

Communication technology brings with it the ability for people to instantly share information with potentially billions of others. This interconnectedness means an even greater exchange of information—a greater marketplace of ideas. Just as oil fuels a fire, this abundance of information will fuel democratic change across the world. Of course, not everyone is in favor of change and governments will seek to prevent information from being exchanged. The same technology that makes it possible to instantly communicate with so many people also makes it possible for governments to try to silence people. Authoritarian regimes will continue to try to shut off the means by which people communicate. Smashing printing presses has evolved into shutting down social media, turning off communication networks, and otherwise using government power to try to eliminate the means of communication.

Ultimately, there will be technical solutions to the government's use of technology to suppress speech—"always bet on the hackers" is a wise adage. For example, although Turkey tried to cut off access to Twitter and YouTube, reports indicate that almost immediately people found ways around the ban. In fact, the attempt to stifle speech resulted in even more discourse about the ban. "After Turkey's Twitter was apparently disabled, the hashtag #TwitterisblockedinTurkey went supernova" (McCoy, 2014).

Although there will be technical solutions there should be legal ones as well. The courts are the protectors of the Constitution and it is up to them to apply the freedoms of speech and association to these new methods of communication. The common law system is accustomed to applying law to new situations and will eventually find the right path. The courts will need to be guided by vigilant protectors of free speech who are familiar with the digital generation's methods of communication. Digital natives will need to make their voices heard though *amicus curiae* briefs such as have been filed by the ACLU, Facebook, and the Electronic Frontier Foundation in recent cases.

In the case of *access to communications,* the court should recognize that when the government seeks to prevent speech by cutting off the means by which people communicate it is intruding up the free speech rights of the people. The BARTs of the world should not feel unencumbered by the Constitution when they choose to silence those with whom they disagree. The FCC and state governments must step up and fill the current vacuum in law regarding wireless interruptions. The US is not an authoritarian regime and it is the legal system that assures this fact. Rather than waiting on a case, the legislative branches and administrative agencies should institute rules that protect speech while allowing for the protection of the public. Yelling fire in a crowded theater should be prohibited. Organizing a protest should not.

Finally, as we invent more ways to communicate the courts will need to understand and embrace these new technologies, at least enough to apply constitutional protections when warranted. Rather than ignore or dismiss modern ways of communication the courts need to listen to the digital natives and tailor the existing legal principles to the new realities. Speech takes many forms and will take many more forms yet unimaginable. The common law is equipped to deal with these changes as long as those in the system approach the task with open eyes.

## REFERENCES

Arsu, S. (2014, March 20). Turkish Officials Block Twitter in Leak Inquiry. *The New York Times*. Retrieved from http://www.nytimes.com

*Bantam Books, Inc. v. Sullivan*, 373 U.S. 58, 70 (1963).

Bay Area Rapid Transit. (2011, December 1). *Extraordinary circumstances only for cell phone interruptions*. Retrieved from https://www.bart.gov/news/articles/2011/news20111201

*Bobby Bland, Daniel Ray Carter, JR., David W. Dixon, Robert W. McCoy, John C. Sandhofer, and Debra H. Woodard v. B.J. Roberts*, 2012, 857 F. Supp. 2d 599 (E.D. Va).

British Broadcasting Corporation. (2011, January 28). *Egypt severs internet connection amid growing unrest.* Retrieved from https://www.bbc.com/news/technology-12306041

Bulwa, D. Wyatt Buchanan and Matthew Yi (2009, January 15), *Behind murder charge against ex-BART officer.* SFGate, Retrieved from http://www.sfgate.com/bayarea/article/Behind-murder-charge-against-ex-BART-officer-3254683.php)

*Clark v. Community for Creative Non-Violence*, 468 U.S. 288, 104 S. Ct. 3065, 82 L. Ed. 2d 221 (1984).

*Cohen v. California*, 403 U.S. 15 (1971).

*Davidson v. Plowman* 715 Fed. Appx. 298 (4th Cir. 2018).

Facebook, 2013, Brief of Facebook, inc. As *amicus curiae* in support of Plaintiff-Appellant Daniel Ray Carter, Jr. And in support of *vacatur*, in the United States Court of Appeals for the Fourth Circuit, Bobby Bland, Daniel Ray Carter, Jr., David w. Dixon, Robert W. Mccoy, John C. Sandhofer, and Debra H. Woodward, Plaintiffs-Appellants, v. B.J. Roberts, Defendant-Appellee.

*Freedman v. Maryland*, 380 U.S. 51 (1965).

Google Glass. (2014, April 6). Retrieved from http://www.google.com/glass/start/

Hidy, K. (2019). Social Media Use and Viewpoint Discrimination: A First Amendment Judicial Tightrope Walk with Rights and Risks Hanging in the Balance. *Marq. L. Rev., 102*, 1045-1085.

Howard, Agarwal, & Hussain. (2011, October). *The Dictators' Digital Dilemma: When Do States Disconnect Their Digital Networks?* Brookings. www.brookings.edu/papers/2011/10_dictators_digital_network.aspx

International Telecommunications Union. (n.d.). http://www.itu.int/en/ITU-D/Statistics/Pages/stat/default.aspx visited 3/25/2014

*Knight First Amendment Institute at Columbia University v. Trump*, 928 F.3d 226 (2nd Cir. 2019).

Leonard, J. (2010, July 8). Former BART officer convicted of involuntary manslaughter. *The LA Times*. Retrieved from http://articles.latimes.com/2010/jul/08/local/la-me-bart-verdict-20100709

McCoy, T. (2014, March 21). Turkey bans Twitter — and Twitter explodes. *The Washington Post*. Retrieved from http://www.washingtonpost.com/news/morning-mix/wp/2014/03/21/turkey-bans-twitter-and-twitter-explodes/?tid=pm_pop

*McIntyre v. Ohio Elections Com'n*, 514 U.S. 334, 115 S. Ct. 1511, 131 L. Ed. 2d 426 (1995).

Miller, G., O'Keefe, E., & Goldman, A. (2014, March 11). Feinstein: CIA searched Intelligence Committee computers. *Washington Post.* Retrieved from http://www.washingtonpost.com/world/national-security/feinstein-cia-searched-intelligence-committee-computers/2014/03/11/982cbc2c-a923-11e3-8599-ce7295b6851c_story.html

*Neb. Press Ass'n v. Stuart*, 427 U.S. 539, 559 (1976).

No Justice No B. A. R. T. (2014, April 6). http://nojusticenobart.blogspot.com/2009/01/what-is-no-justice-no-bart.html

Ostrove, E. & Leber, Y. (2019, April) Silencing Critics in the Age of America's First Tweeter-In-Chief, Where Social Media Meets the First Amendment. *New Jersey Lawyer, The Magazine*, 317.

Policinski, G. (2011, August 19). BART Cell Shutdown a Landmark in Cyber-Assembly. *First Amendment Center.* http://www.firstamendmentcenter.org/bart-cell-shutdown-a-landmark-in-cyber-assembly

Protecting Cyberspace as a National Asset Act of 2010, S.3480 - 111th Congress. (2010).

*Roth v. U.S.*, 354 U.S. 476, 77 S. Ct. 1304, 1 L. Ed. 2D 1498 (1957).

Shakespeare, W. (1598). *Henry IV.* London: Stationers Company.

Siy, Feld, Bankston, Llansó, & Tien. (2012, April 30). *Comments Of Public Knowledge.* Center For Democracy & Technology, Electronic Frontier Foundation, Benton Foundation, Free Press, Minority Media And Telecommunications Council, National Hispanic Media Coalition, And Open Technology Institute At The New America Foundation.

*Snyder v. Phelps*, 131 S. Ct. 1207 (2011).

*Turner Broadcasting System, Inc. v. Federal Communications Commission*, 512 U.S. 622, 114 S. Ct. 2445, 129 L. Ed. 2d 497 (1994).

Twitter.com. (n.d.a). *New user FAQs.* https://support.twitter.com/groups/50-welcome-to-twitter/topics/203-faqs/articles/13920-new-user-faqs

U.S. Const. amend. I.

Twitter.com. (n.d.b). *Help Center.* https://help.twitter.com/en/using-twitter/blocking-and-unblocking-accounts

*United States v. Wei Sang Phua*, 100 F.Supp.3d 1040 (2015).

*Wallace v. Jaffree*, 472 U.S. 38, 105 S. Ct. 2479, 86 L. Ed. 2d 29, 25 Ed. Law Rep. 39 (1985).

Williams, E. (2019), First Amendment Protection Against Curtailment of Access to, or Retaliation for Communications on, Social Media. *American Law Reports*, 38 A.L.R. Fed. 3d Art. 5 Sec. 5.

## ADDITIONAL READING

Balkin, J. (2004). Digital Speech and Democratic Culture: A Theory of Freedom of Expression for the Information Society. *N.Y.U. Law Review*, *79*, 1–55.

Balkin, J. (2009). The Future of Free Expression in a Digital Age. Faculty Scholarship Series. Paper 223. http://digitalcommons.law.yale.edu/fss_papers/223

Berman, J., & Weitzner, D. (1995). Abundance and User Control: Renewing the Democratic Heart of the First Amendment in the Age of Interactive Media. *The Yale Law Journal*, *104*(7), 1619–1637. doi:10.2307/797026

Bollinger Lee, C. (2010). *Uninhibited, Robust, And Wide-Open a free press for a new century*. Oxford: Oxford University Press.

CALIFORNIA LEGISLATURE— 2011–2012 REGULAR SESSION SENATE BILL No. 1160.

China Law & Practice. (2011). Ministry of Culture, Tentative Provisions for the Administration of the Culture of the Internet.

Coleman, G. (2009). Code is Speech: Legal Tinkering, Expertise, and Protest among Free and Open Source Software Developers. *Cultural Anthropology*, *24*(3), 420–454. doi:10.1111/j.1548-1360.2009.01036.x

Comninos, A. (2011). E-revolutions and cyber crackdowns: User-generated content and social networking in protests in MENA and beyond. Retrieved from http://giswatch.org/sites/default/files/gisw_-_e-revolutions_and_cyber_crackdowns_0.pdf

Comninos, A. (2012). *Freedom of Peaceful Assembly and Freedom of Association and the Internet*. Commissioned by the Association for Progressive Communications.

Dowell, J. (1998). Bytes and Pieces: Fragmented Copies, Licensing, and Fair Use in a Digital World. *California Law Review*, *86*(4), 843–877. doi:10.2307/3481141

Fleming, J. (2012). The Case for a Modern Public Forum: How the Bay Area Rapid Transit System's Wireless Shutdown Strangled Free Speech Rights. *Washburn Law Journal*, *51*(631).

Hunter, D. (2003). Cyberspace as Place and the Tragedy of the Digital Anticommons. *California Law Review*, *91*(2), 439–519. doi:10.2307/3481336

International Telecommunication Union. (1989). Final Acts of the World Administrative Telegraph and Telephone Conference.

Kenyon, A. (2010). What Conversation? Free Speech and Defamation Law. *The Modern Law Review*, *73*(5), 697–720. doi:10.1111/j.1468-2230.2010.00815.x

La Rue, F. (2011). *Report of the Special Rapporteur on the promotion and protection of the right to freedom of opinion and expression*. United Nations General Assembly.

Lemley, M., & Reese, A. (2004). Reducing Digital Copyright Infringement without Restricting Innovation. *Stanford Law Review*, *56*(6), 1345–1434.

Lucchi, N. (2011). Access to Network Services and Protection of Constitutional Rights: Recognizing the Essential Role of Internet Access for the Freedom of Expression. *New Europe Law Review*, *19*(3), 645–678.

Lunney, G. (2001). The Death of Copyright: Digital Technology, Private Copying, and the Digital Millennium Copyright Act. *Virginia Law Review*, *87*(5), 813–920. doi:10.2307/1073857

*Meyer, Colorado secretary of state, et al. v. Grant et al*, 486 U.S. 414; 108 S. Ct. 1886; 100 L. Ed. 2d 425 (1988).

BBC Monitoring Trans Caucasus Unit. (2012). Iran government-sponsored internet to cut off people from global web – sites.

O'Brien, K. (2012, June 8). E.U. seeks clarity on access to mobile Web; Regulators and industry debate disclosure rules and minimum speeds. *International Herald Tribune*.

Olson, D. (2010). First Amendment Based Copyright Misuse. *William and Mary Law Review*, *52*(537).

Price, M. (1995). Free Expression and Digital Dreams: The Open and Closed Terrain of Speech. *Critical Inquiry*, *22*(1), 64–89. doi:10.1086/448782

Reisinger, D. (2009, October 4). Finland makes 1Mb broadband access a legal right.

Richtel, M. (2011, January 28). Egypt Cuts Off Most Internet and Cell Service. *The New York Times,* pp. A13.

Robbins, I. (2013) What Is the Meaning of "like"?: The First Amendment Implications of Social-Media Expression. *Federal Courts Law Review, 127.*

Rutzen, D., & Zenn, J. (2011). Association and Assembly in the Digital Age. *The International Journal of Not-for-Profit Law*, *13*(4).

Seldin, J. (2012, November 29). Government Cuts Internet Access Across Syria.

Sklan, A. (2013). @SOCIALMEDIA: SPEECH WITH A CLICK OF A BUTTON? #SOCIALSHARINGBUTTONS<> *Cardozo Arts & Entertainment Law Journal, 32.*

State of Arizona House Bill 2549. (2012) AMENDING SECTIONS 13-2916 AND 13-2923, ARIZONA REVISED STATUTES; RELATING TO ELECTRONIC OR DIGITAL DEVICES.

Vreese, C. (2007). Digital Renaissance: Young Consumer and Citizen? *The Annals of the American Academy of Political and Social Science*, *611*(1), 207–216. doi:10.1177/0002716206298521

Weitzner, D. (2007). Free Speech and Child Protection on the Web. *IEEE Internet Computer Technology & Society column*. Retrieved from http://dig.csail.mit.edu/2007/06/ic-labeling-free-speech-weitzner.html

Wimmer, K. (2006). Toward a World Rule of Law: Freedom of Expression. *The Annals of the American Academy of Political and Social Science*, *603*(1), 202–216. doi:10.1177/0002716205282120

## KEY TERMS AND DEFINITIONS

**Facebook "Like":** On the social media website Facebook, users have the option to click a thumbs up icon and "like" any post or page. When they do this, they are signaling their support or interest in the subject. For example, one might "like" the page of a favorite candidate or a funny video of a cat. When one "likes" a page this information is recorded as support and the page appears in the person's own timeline. This act of liking is speech in itself.

**Marketplace of Ideas:** This is a concept relating free speech and democracy. The idea is that lots of different opinions and facts will be shared in a virtual marketplace. The best ideas will come forward and become policy. The best marketplace is one where all actors have free access to express themselves. Only through this free exchange of ideas will the best ones mature into policy.

**Prior Restraint:** This is a type of restriction on speech that occurs before the speech itself happens. For example, the government seeks to prevent publication of an article exposing government corruption. Since this is an attempt to prevent speech from happening courts apply particular scrutiny to these situations.

**Speech:** Under the law, speech is much more than words that emanate from one's mouth. Written words, symbolic expressions, sit ins, wearing arm bands, burning flags, and liking a page on Facebook all have been found to be speech under the law. This encompassing idea of speech is important because the law extends constitutional protections only when something is characterized as speech.

**Time, Place, and Manner Restriction:** The courts distinguish between content-based restrictions and neutral restrictions. Content based treats different content differently. For example, banning graphic nudity in public areas is an example of a content-based restriction. A neutral restriction is one that treats all speech equally regardless of the content. Time, place, and manner restrictions are examples of content neutral regulations. For example, a state may ban billboards over a certain size to protect the aesthetic beauty of the land. This is an example of a neutral restriction on the manner of speech. Courts apply a much lower standard when examining a neutral time, place, or manner restriction then when examining a content-based restriction.

# Chapter 5
# Online Resolution and Citizen Empowerment:
## Tax Appeals and Court Resolutions in North America

**Colin Rule**
*Santa Clara University, USA*

**Mark James Wilson**
*Tyler Technologies, USA*

## ABSTRACT

*In the private sector, the growth in interactive, online technology use has already disrupted many private industries, from medicine to finance to entertainment. Interactive, online technology has empowered consumers, giving them more choices and better information, which has in turn transformed global business. Now it is clear that government services are in the process of a similar transformation. Government agencies face unique challenges in implementing interactive, online technology, and understanding best practices can be a challenge. In this chapter, the authors describe their first-hand experience helping government agencies build advanced online dispute resolution systems. They focus in particular on two case studies: transitioning property tax appeals from a paper-based process to an interactive online process and introducing interactive online technology into the courts. Through this examination, they (1) highlight the unique challenges we encountered and (2) make recommendations for government agency decision makers from the lessons we learned.*

DOI: 10.4018/978-1-7998-3152-5.ch005

## INTRODUCTION

In the United States, individual citizens now use information and communications technologies routinely in many areas of their lives. According to the Pew Research Center's Internet & American Life Project (n.d.), 98% of Americans ages 18-29, 92% of Americans ages 30-49, and 83% of Americans ages 50-64 use the internet. In the private sector, this growth in interactive, online technology has already disrupted many industries, from medicine to finance to entertainment. Technology has empowered consumers, giving them more choices and better information, which has in turn transformed business. Now it is clear that government services are in the process of a similar transformation. This individual empowerment is fundamentally changing the relationship between citizens and government, and the significance of this change may yet turn out to be more revolutionary in the public sphere than it has been in the private one.

Most individuals are now quite familiar with the interactive, online technology model of logging into a service provider's website, picking a username and password, and filling out forms and uploading information. Whenever a question arises about the service in question, the user can log back into the website, check the status of their request or case, post new information, and log out again. This kind of asynchronous, always-on interface is commonplace for people trying to make health care elections for their private insurance plan, or rebalance their 401-K with their bank, or sign their kids up for summer camp. It is also familiar to the billions of people who have used eBay, Facebook, LinkedIn, or any number of online services. But it is still uncommon in government services.

Public agencies move slowly, by design. Investment in public infrastructure is much more deliberate than the private sector. Government rarely is the first adopter of new innovations. But once an innovation has proven its utility, government investments in that innovation can be sustained and long-term. As the internet becomes more ubiquitous, the value of building the future on interactive online technology is beginning to make more sense to public decision makers.

In this chapter, we discuss the efforts of public agencies to implement interactive, online technology to support their work, focused particularly on two case studies we have come to know in depth: property tax assessment appeals and civil claims in courts. First, we discuss the challenges faced by public agencies, and how citizen expectations have shifted over the past decade. Second, we present some background about cloud-based models for software development, and the benefits they can bring to public agencies looking to meet changing public expectations. Third, we examine the property tax appeals process, focusing in particular on efforts to make appeals more interactive and efficient through the use of interactive, online technologies. Fourth, we examine efforts in courts to empower citizens to resolve their own disputes without requiring a hearing. Fifth and finally, we generalize some of the lessons we have learned in working with governments, and distill a set of recommended best practices that can help guide public decision makers as they contemplate how to update their services for a networked world.

## CHALLENGES FACED BY PUBLIC AGENCIES IN UTILIZING INTERACTIVE, ONLINE TECHNOLOGY

Ten years ago, proposals to move government services onto the web were perceived in many circles as elitist. Because computers and internet connections were expensive, it was thought that spending public resources on internet projects would benefit only those affluent enough to afford access. This "digital divide" was a compelling enough concern to delay investment in internet-based service channels for many public agencies. But as the cost of access has come down, and internet access has become more ubiquitous (particularly through mobile devices), concerns about the digital divide have faded.

But even with the easing in concerns around the digital divide, many government agencies have still not yet implemented interactive, online interfaces for their citizens. For many local, state, and federal government agencies, forms must still be submitted in person, by mail, or by fax. If online filing forms are available, they are quite simple; submitted information is simply forwarded to an agency employee's email inbox, which means the submitter cannot log back into a website and see updated status information for their submitted request. These approaches offer little of the power available through common websites like Amazon and Facebook, which count a majority of Americans as users.

Public agency employees are often painfully aware of the shortcomings of their existing technologies, and given no resource constraints, they would happily install updated platforms that provide more sophisticated interactive, online interfaces. Unfortunately, there are several types of obstacles faced by these agencies in making these updates. Some of the most common include:

- **Expense:** Technology systems are often expensive to implement, because the administrative processes they must support are complex. There may be dozens of case types and hundreds of statuses for individual cases, and hundreds if not thousands of individual agents working the cases in question. Implementing technology that can capture all of that complexity is no simple challenge.
- **Legislative requirements:** There are often crisscrossing laws that govern how public agencies can do their work. Valid concerns about privacy and

information security have led to very restrictive rules around how agencies can store and share data, some of which were put in place long ago, before technology became more mature and flexible. This makes creating interactive, asynchronous information portals that comply with existing law extremely complicated, and creates incentives for public sector leaders to wait until new legislation is implemented that may offer more flexibility.

- **Unique Processes:** Each public agency has usually developed its own particular process design. No two agencies are identical in their case management or interaction processes. Each has unique staff roles, filing forms, and rules and policies. Rainey and Katsh (2012) stated, "When new online technology is created for any process, the initial impulse is to create online mirror images of the 'live' or offline process" (p. 248). Some agencies aim to replicate exactly their current processes online. Public agency staff may have been using the existing system for so long that it may be difficult for them to envision the new system as something other than an online replica of their offline process.
- **Long Timeframes:** In addition, government contracting can take a lot of time. Contracts usually require many levels of review by legal and even governing bodies. Many public agencies require the issuance of a request for proposal (RFP) for new technology. This requires time for the public agency to issue the RFP, the time for a response, and the time for a determination.

All of these challenges must be taken into consideration when evaluating the decision to upgrade technology. Many agencies invested heavily in mainframe technology more than a decade ago to power their operations, and it has proven extremely costly to keep that technology up to date. It is possible to build services on top of these older mainframes, but it can be very expensive and time consuming. In light of all the "sunk costs" public agencies have put into their current information architecture, the decision to upgrade to a new technology – which represents significant implementation risk in addition to the cost of abandoning existing infrastructure – is not a simple matter.

## SHIFTING USER PREFERENCES

Tolbert, Mossberger, and McNeal (2008) defined e-Government as "the delivery of [government] information and services via the Internet or other digital means…e-Government is now one of the fastest-growing activities online" (p. 549). It is striking how quickly citizen preferences have changed. In a matter of years the way citizens want to interact with their government has shifted from in-person and over the telephone to online connectivity. The Pew Center for Internet and Society (2010) recently completed a study called *Government online: The internet gives citizens new paths to government services and information*, which reported the following:

- "48% of internet users have looked for information about a public policy or issue online with their local, state or federal government" (p. 2).
- "Most online government users engage in a range of government website transactions…69% of these online government users do at least three different activities" (p. 17).

For many years the conventional wisdom was that citizens wanted to engage with their government agencies in person or by voice, to interact on a human level. The thought was that such exchanges would foster trust and connection by humanizing public employees, breaking the veneer of the aloof and disconnected government agency. However, much like the banking industry, which faced similar skepticism with the roll out of ATMs in the 1980s, citizens now value efficiency over human engagement, and user preferences have shifted to online channels for that reason. Tessler (2010) reported, "…the Pew Internet & American Life Project found that 82 percent of Internet users -- or 61 percent of American adults -- had looked up information or completed a transaction on a government Web site over the previous year" (para. 2).

The government of British Columbia (the province sometimes referred to as the "Silicon Valley of Canada") has been one of the most innovative adopters of information technology (IT) in public service. Their Ministry of Justice recently created an online Civil Resolutions Tribunal to move low value civil cases into online processes, making them one of the first government agencies in North America to do make an online process mandatory (British Columbia, 2012). Recently they conducted a series of surveys of their citizens asking how they would most prefer to access public services, and the results were quite striking. A selection of the survey results are presented in Figures 1-3 (2010 B.C. Citizen Satisfaction with Provincial Government Services report, 2011).

Figure 1 analyzes the channels citizens are using for their first engagement with a public agency, and which they use as their principal channel. This chart makes clear that the internet is now the dominant first channel for most citizens looking to engage with a public agency. When a citizen has a question, issue, or concern, the majority of the time their first step is to go to the internet to try to find the answer. Almost twice as many respondents used the internet as their first channel compared to the second channel, which was in-person at a service counter. For subsequent communications, other channels made a stronger showing – in particular, in-person at service counters and the telephone. But still, the internet is the clear preference in subsequent engagements as well. Also, it should be noted, the "internet" in this context means online interactive services or web sites, because email-only interaction was also called out separately in the survey.

Figure 2 focuses on direct contacts into public services, meaning interaction with agency staff as opposed to information-gathering from static web pages. For first human contact, the majority of respondents did indicate that they still make use of in-person channels. Maybe if a citizen cannot answer their questions from the information on the agency website, or if they are unsure what processes are available or what their rights may be, an in-person interaction can help to orient them to the appropriate process and set them on the right path. Mail and the internet are in second and third places when it comes to first contact, probably because responses can take longer and are usually more generic. For the second contact, the telephone and internet emerge as the clear preference, probably because the benefit gleaned from the first communication in terms of process orientation and expectation setting has already been achieved. By the third contact, and for all future contacts, the internet is the clear preference. At this stage, orientation is no longer necessary, and efficiency and convenience are prized above all.

Figure 3 may be the most surprising of the three charts. It shows satisfaction rates for the various communications channels, paired with preference metrics. This chart makes clear that the internet does not drive the highest satisfaction of all the available channels. In fact, it places fourth. Email, postal mail, and in-person interaction at service counters all rank higher in terms of citizen satisfaction, with

*Figure 1. Frequency of use of interaction types as first/principal channels in British Columbia (British Columbia, 2011)*

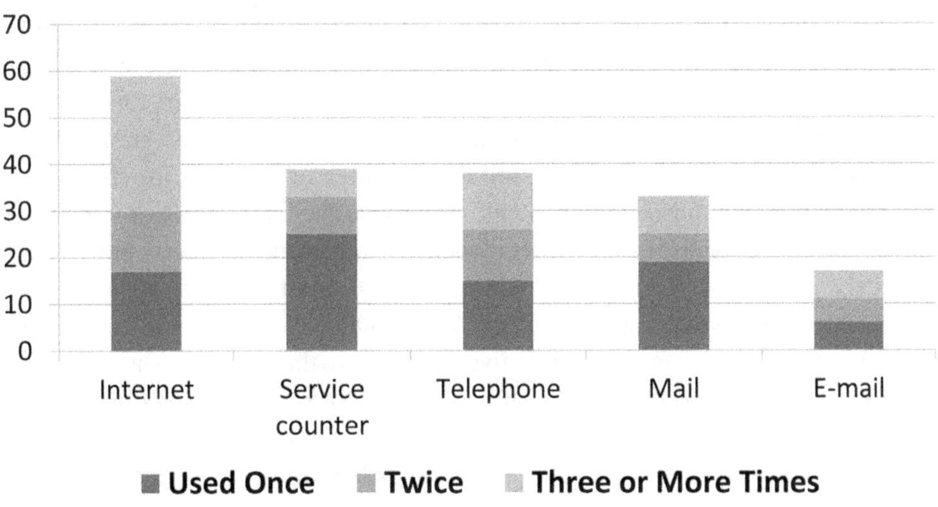

*Figure 2. Frequency of use of various public interaction channels in British Columbia (British Columbia, 2011)*

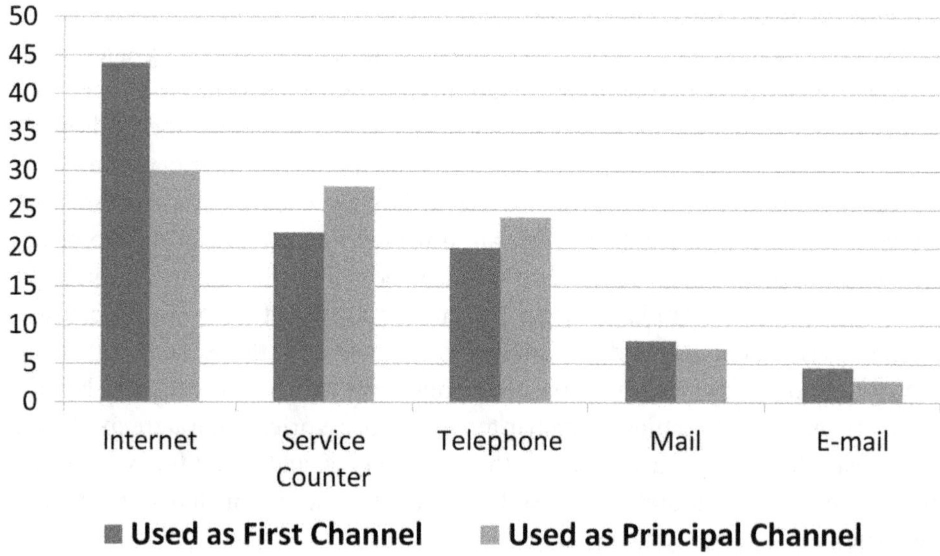


in-person interaction generating the highest satisfaction numbers. This can be seen as a validation for those who have argued that in-person interaction best establishes the connection between citizens and their government, and builds trust. However, even in light of those satisfaction numbers, the internet is still the preferred channel for citizen interaction with public services.

Yes, our satisfaction may be higher when we can interact with a friendly, smiling person at a service desk and have my questions answered quickly and effectively. But we still prefer being able to log into a website at midnight in our pajamas because it saves us having to take time off work to drive down to the government office. For us, and many others like us, the efficiency benefit outweighs the satisfaction improvement. Still, the lower satisfaction rates associated with the internet channel may have more to do with the current state and effectiveness of the properties of the internet channel, and may very well be indicative of a calling for changes to this channel. In its "Introduction to Online Service Delivery," British Columbia (2010 B.C. Citizen Satisfaction with Provincial Government Services report, 2011) reported, "Governments around the world are reinventing the way they provide information and services to their citizens. As demand for online services increases, our web properties must evolve to effectively respond to the needs of citizens" (p. 5).

There is good reason to believe that these results from British Columbia are representative of the broader preferences across North America as a whole. In the years since these data were gathered, these trends have only intensified. Based on

*Figure 3. Frequency of use of various public interaction channels in British Columbia (British Columbia, 2011)*

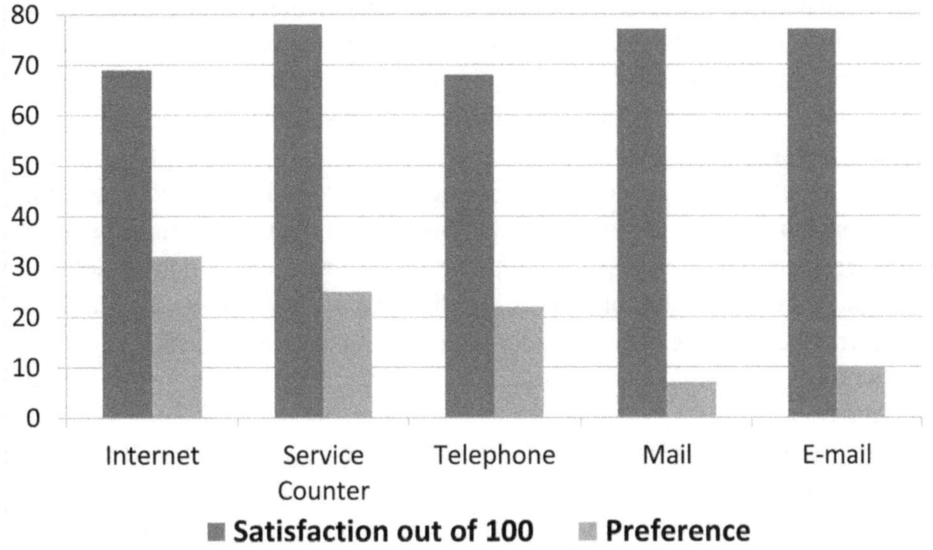


our experience working across the U.S. and Canada, we believe that similar surveys would likely generate similar results in most areas in the United States.

## INTRODUCING "THE CLOUD" TO PUBLIC AGENCIES

Any interaction with public media about technology these days will likely include some reference to "the cloud." While understanding of that term may be slightly different across various sectors of society, the broader concept has begun to take root. And it is true that in Silicon Valley, the preferred model for modern internet service delivery is cloud-based.

Most of us use some cloud-based service every day. Many of the most popular email service providers are cloud based, from Hotmail to AOL to Yahoo to Gmail. Facebook and LinkedIn are cloud based. eBay and Amazon are cloud based. (In fact, Amazon has extended their cloud architecture for others to use, and now companies from Netflix to Dropbox are partially hosted in the Amazon Cloud.) Some leading companies, such as Salesforce, have long promoted the fact that they live entirely in the cloud. Even venerable offline software suites, like Microsoft Office and Adobe Photoshop, have migrated their services to the cloud.

But what exactly is the cloud? At the simplest level, the cloud is where your data and your services are located if they aren't on a hard drive on your local computer.

Instead, they live (or are "hosted") on a server somewhere out in "the cloud," waiting for you to access them through the internet. All you need is a simple display program (like a web browser or app) on your laptop or mobile phone to connect through the network to your data and services, which may in fact be residing in Colorado or Calgary or Chennai –or possibly residing in all of those places at once, because your information can be mirrored across multiple locations so you can get at it quickly no matter where you may be.

Traditionally public agencies have issued complex RFPs to select a software developer who can build a massive software suite to run all of the internal operations of the agency. Figure 4 represents that traditional software development model. In this architecture, all of the software is installed and maintained locally. There is a clear barrier between the software and the internet, and no external users are allowed into the local software. Data and functionality are maintained in local hardware configurations.

Building software in the cloud works quite differently. Software and hardware are accessed via the internet, meaning computers are just the end terminals required to display the information coming in over the web. Functionality is delivered from a remote server, so it doesn't matter if the end user is on an Apple laptop or a Windows desktop or a Linux tablet or an Android phone. All that is required is that the end

*Figure 4. The usual design of a locally hosted, on-site software platform*

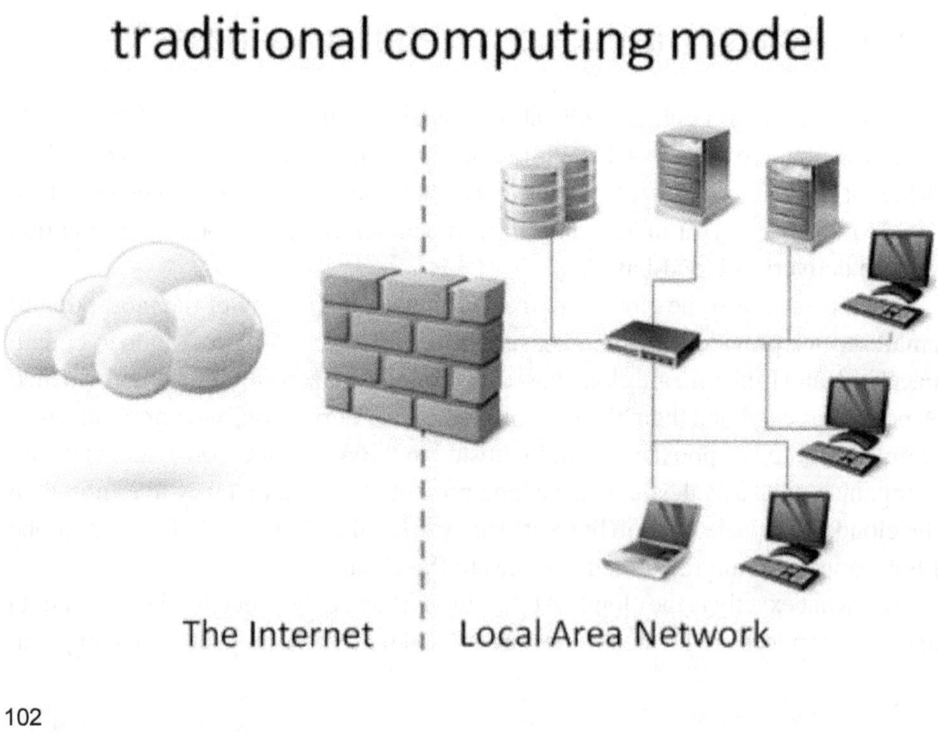

user can see the web page generated by the remote application. This design is often referred to as "Software as a Service" or a "SaaS" model, because instead of being paid to design and deliver a large software platform to run locally, the software designer takes on the continuing responsibility to maintain and upgrade the system running in the cloud. Figure 5 represents the SaaS/cloud development model. This creates an ongoing service relationship and partnership with the client, instead of delivering a software program to the required specifications and then handing administrative responsibilities to the agency.

The benefits of SaaS/cloud based designs are clear. First, they are often much less expensive than traditional development. Because the service provider hosts all the software code on external servers, they can easily re-use components to build new processes much more quickly. This also reduces time to launch, so systems that in the past took years to build can be up and running in months. Because the software is maintained centrally, it can be patched and upgraded constantly, which can be done in such a fashion that is almost seamless to the end users. Software is much less buggy because all of the services are based on the same core platform, instead of dozens or hundreds of different installations all hosted locally with slight variations in each implementation. When a new feature is added to the platform,

*Figure 5. The design of a cloud-based, software-as-a-Service (SaaS) platform*

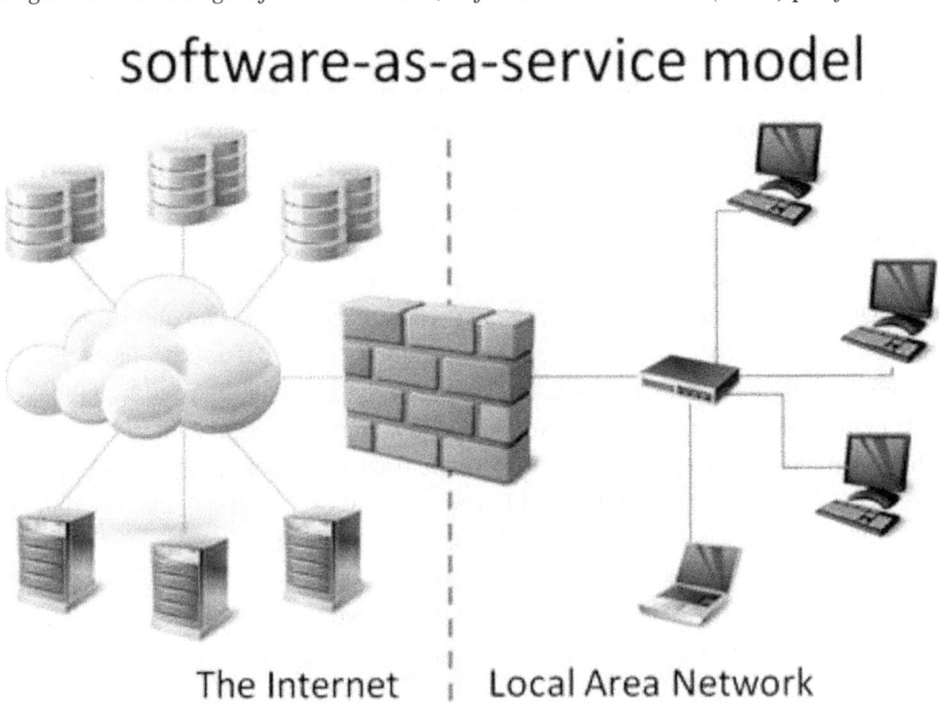

all customers get it quickly. Finally, because there is redundancy across multiple servers, the system is highly stable and highly available, because if one server goes down or becomes inaccessible, others immediately step in to instantly compensate.

All that said, even though cloud based architecture is becoming the dominant standard in Silicon Valley, these types of designs can still generate concerns for many public agencies. One oft-stated worry involves having sensitive data hosted outside of the agency's IT department. The fear is that cloud based systems may be more vulnerable to hackers because the data is not closely controlled by internal IT staff. However, recent studies have shown that data hosted in the cloud is even more secure than locally hosted data. As industry expert Derek Brink (2010) recently explained:

*Drawing on the findings from multiple benchmark studies on best practices in content security and security software as a service, Aberdeen's analysis shows that users of cloud-based web security had substantially better results than users of on-premise web security implementations in the critical areas of security, compliance, reliability and cost. Compared to companies using on premise web security solutions, users of cloud-based web security solutions had 58% fewer malware incidents over the last 12 months, 93% fewer audit deficiencies, 45% less security-related downtime, and 45% fewer incidents of data loss or data exposure (para. 1).*

In fact, a primary benefit to public agencies of cloud-based SaaS architectures is that information intended to be accessible to outside users can be hosted offsite, while sensitive internal information may continue to reside only on servers inside the agency's four walls. Hosting services for outside users on hardware inside the agency may increase the risk of data vulnerability, while hosting services for citizens outside the agency can help to protect internal systems from unauthorized data access. Then the inside and outside systems can communicate securely to keep information up to date.

It can be enormously expensive and complicated to build software internally, hire technology experts to maintain the software, and to keep it up and running at the standard now expected by online users (usually five nines of uptime, or 99.999% availability). Even if a system works flawlessly at launch, it can be a challenge to keep it constantly patched, running smoothly, and fully redundant. The reason why cloud-based models and SaaS implementations are becoming more popular is because all of these tasks can be handled by external organizations that specialize in handling them for multiple clients, freeing resources within agencies to focus on tasks more relevant to the agency's core responsibilities.

All of this talk about technology at a holistic level can get a little abstract, so let's now focus our discussion on two case studies within public agencies that are going through this transition at this very moment: property tax assessment appeals and court civil cases.

## HOW PROPERTY TAX ASSESSMENT APPEALS WORK

Homeowners in the U.S. are all familiar with the process of receiving a property tax bill every year, based on the assessed value of their house. Taxes are levied against almost all properties across America, whether commercial, industrial, or residential. Some properties are classified as exempt from taxation based on various state specific factors; these exemptions are generally granted by the taxation agency and could be the subject of disagreement or appeal. Taxation of properties can extend beyond real estate and buildings to personal property.

The International Association of Assessing Officers (IAAO) (2010) explains property tax and its role in funding local government:

*The property tax provides for balance and equity in the total tax system by taxing the one element of ability to pay overlooked by other state and local taxes. The property tax allocates the cost of government according to ability to pay as measured by property wealth. Among the many types of taxes levied, the property tax is the only tax used in every state of the United States, the District of Columbia, and every Canadian province. In fact, the property tax remains the most important source of own-source and total revenue for local governments in the United States (p. 6).*

The formula to compute property tax varies from county to county, but in most cases the local property tax assessor estimates the market value of your house and sends a tax bill that is a percentage of that value. Some counties have laws that cap the rate of growth of property taxes, so even if your house increases 10% in value in one year, for example, your property taxes can only go up 1% or 2% a year. But property taxes are not only for residential properties. Every commercial property also pays taxes, so for large buildings (e.g. hotels, office buildings) the tax bill can be quite large.

Property Tax Assessors make use of advanced software, called CAMA (Computer Assisted Mass Appraisal) software, to help calculate and to track the values of every parcel in their district and send out all the property tax bills. This software uses a combination of human powered and mathematical algorithms to estimate the accurate property tax values at any given time. This is difficult to do well, because valuations are constantly in flux as a result of economic changes and factors like mortgage rates and overall market supply and demand. And getting out all the tax bills on time can be a huge challenge, particularly in counties with hundreds of thousands or millions of parcels within their jurisdiction.

There are dozens of major suppliers for CAMA platforms, from market leaders like Tyler Technologies and Thomson-Reuters/Manatron, to smaller regional systems supported by local software developers. Large counties can have very large property tax bases, which justify significant expenditure on the part of the counties to purchase advanced software. But the complexity of installing a large CAMA system in very large counties can be overwhelming. Once a county commits to using a CAMA system the transition can take years to fully complete. And once a county is migrated onto a CAMA system, it is quite committed to continue using that system, because by law the property tax bills must come out on a certain date every year, come hell or high water. The risk of changing CAMA platforms is therefore high.

While these CAMA systems are quite advanced in the areas of managing property tax valuations and sending out bills, one area they have traditionally not focused on is the area of appeals. By law, every taxpayer has the right to appeal their property tax bill if they feel the amount is inaccurate. The IAAO (2013) described the assessment appeals process as

*...an important component in the assessment process. Appeals provide an opportunity for property owners to meet with the assessor to inquire about their assessments and to learn about assessment and appeal procedures. In the case of disputes about assessments, an appeal system should provide opportunities for both informal meetings with the assessor and formal hearings before independent bodies to resolve disputed issues and thus assure the public that assessments are correct, fair, and equitable. Key to any assessment appeal system is an open and transparent process that relies on a clearly written set of procedures and provides due process (p. 5).*

Taxpayers are quite sensitive to their assessed values, and are quick to file an appeal if they feel the valuation is too high or if they see fluctuations in their valuation. Tergesen (2007) reported:

*The housing market may have softened, but many homeowners are discovering one place where their property values remain at peak levels: on local property-tax rolls. 'They were valuing our home for way more than it's worth,' says Cara Reeves, who recently saw the assessment on her three- bedroom home in Cincinnati jump from $251,000 to $334,400, pushing her latest property-tax bill up 82%... Because many local governments assign values to properties in their jurisdictions only once every three or more years, many homes still carry assessments from the market's peak, in 2005 and 2006 (para. 1).*

The Assessor often has a defined window when taxpayers are given the right to file a formal appeal of their bill. This window usually begins just after the tax bills are sent out, and closes two or three months later. While the tax bill distribution process is quite automated, this appeals process is usually quite manual. Taxpayers may file their appeal in person, or via the mail, and they may have to fill out a paper form that asks questions about why they feel their assessed value is incorrect. They also may be required to submit evidence bolstering their case that the taxable amount is wrong. In addition, many appeals are filed by agents on behalf of taxpayers, and those agents may be managing hundreds or thousands of appeals at one time. As Tergesen (2007) put it, "Help is available. A mini-industry of consultants, attorneys, and real estate agents has emerged to assist homeowners in preparing appeals" (Do it Yourself section, para. 2). High volume filers have their own needs, different than individual taxpayers filing on their own behalf.

Most assessing jurisdictions follow a very similar appeal pattern. The IAAO (2013) provides a common process framework in its *Standard on Assessment Appeal*:

1. **Assessment Notice:** The assessing jurisdiction notifies the taxpayer of his assessed value: "When an assessment is changed, a notice of assessment that identifies the property, the property owner, the estimated market value, and the assessed value of the property should be mailed to each property owner" (p. 8).
2. **Filing an Appeal:** A filer or his agent fills out required filing forms and provides required evidence directly with the assessing jurisdiction (p. 5).
3. **Informal Review:** The assessing jurisdiction has the first opportunity to review the appeal. According to the IAAO (2013), "The appeal process should begin with an informal consultation between the assessor and the property owner in order to Identify and document errors, Review the equity and uniformity of assessment...Determine what issues (facts) the parties to a valuation dispute can agree on...[and] Identify and clarify the basis for an exemption or assessment limitation claim" (p. 5-6). Oftentimes, a negotiation will occur between the parties and the assessing jurisdiction may offer a value settlement offer. This process may continue until a fair value is reached. If a value cannot be agreed upon, the filer may escalate to a formal review.
4. **Formal Review:** When the dispute cannot be resolved at an informal level, it can be forwarded to a local or regional board "as the first level of formal appeal" (IAAO, 2013, p. 6). This board can be made up of property assessment experts and volunteers within the community whose charge is to maintain fairness in the valuation. The filer usually meets with the board to present its case. The assessing jurisdiction usually acts as the respondent in the matter heard by the board. A filer may have about 15-20 minutes to present its case.

If the board makes a decision that the filer disagrees with, they can generally appeal to a state board.

5. **State Board / Courts:** According to the IAAO (2013), "The state or provincial property tax tribunal should be the final administrative arbiter for individual appeals. However, unresolved legal and appraisal issues may be appealed to the courts. For efficiency, state or provincial property boards may constitute the only level of formal appeal before appeal to the courts" (p. 6).

Because of all these steps, an appeal can take quite a bit of time to move from filing to closure. On average, based on our observations, property tax assessment appeals processes in the United States usually take around 18 months to complete. That means that if a taxpayer gets a bill they think is too high and they immediately file an appeal, they are still required to pay the bill they disagree with that year, and then to pay it again the next year, before a decision is finally rendered. If they win their appeal, the amount is adjusted moving forward, and they receive reimbursements plus interest for the prior two tax cycles they had already paid. But as you might imagine, this long process can generate frustration on the part of the taxpayers.

Our 18-month estimate is an average, so it should be noted that some of these cases go much longer than a year and a half. Much of the time required to resolve these cases is due to administrative tasks. By law, many assessors are required to process an appeal even if the information submitted is very sparse – say, a one line note written on a postcard that says "I want to appeal my property tax bill!" Once an appeal is filed, there is a lot of back and forth between the Assessor and taxpayer, getting the file complete, answering certain required questions, and ensuring all required documents are filed. There may be visual inspection of the property required as part of the appeals process (though sometimes that visual inspection can happen via satellite imagery or street-view image capture, which is much more efficient). If the case is escalated to an evaluative process at the state level, all of the information captured by the Assessor's Office must be transmitted to the new office, where it is reviewed again, which may result in more evidentiary obligations.

Most of these appeals are worked on a first-come, first served basis, so cases that may be relatively simple to work out through mutual agreement have to wait for other cases ahead of them in the queue to be resolved before they get addressed. Also, an appeal may be efficiently progressing through the process, but there are few automatic updates generated by the system to keep the taxpayer informed as to case status. That means that whenever a taxpayer wants an update on the status of their case, they have to call into the Assessor's Office and speak to an employee, who in turn has to check in the internal systems to find out the latest information. That can generate even more delays, as employees in the Assessor's Office have to take time out of their schedule to respond to status queries from taxpayers.

## Case Study #1: Filing an Appeal – Paper Process

In one of thousands of Property Assessor offices across the United States, hundreds of taxpayers crowd and line the halls of city hall—waiting for hours to have their property assessment appeal filed and heard. Sheila, a taxpayer, makes her way up to the appeal counter after waiting 45 minutes to find out that she does not have the documentation required to properly file an appeal. She'll have to return home and come back another day, before the 20-day filing deadline is over. Sheila doesn't know how she'll be able to return - requiring absence from one of her two full-time jobs and requiring a 2-hour bus ride into the center of town. She may just have to bag the appeal and pay the taxes, even though she is almost sure her property has been overvalued.

## BRINGING ASSESSMENT APPEALS INTO THE CLOUD

In the past two years, several counties have begun creating interactive, online interfaces for managing their property tax assessment appeals. These systems offer compelling benefits both to external users (taxpayers and agents) as well as internal agency staff. Rabinovich-Einy and Katsh (2012) stated the following:

*The contemporary ODR [online dispute resolution] landscape can be divided into two principal domains—tools and systems. The first area has involved the development of specific dispute resolution applications that can be used to resolve both online and offline disputes...ODR systems include ODR tools...used in a coordinated way within a closed setting by a limited (but potentially very large) number of users who are engaged in ongoing interactions with other users and may experience similar types of problems over time (p. 40-42).*

Presently, some assessing jurisdictions employ a selection of *tools* (e.g., a web filing form, barcoded documents, a status checking tool), but the vast majority lack any true ODR system.

In an effective ODR system, the filing process for an assessment appeal resides online. Taxpayers can log in any time to fill out the dynamic, step-by-step form, which collects all relevant information about the appeal. The form can also serve to educate the taxpayer about how their property was valued, and how the tax amount was calculated. For example, the system may automatically share recent sales prices for comparable properties close by the taxpayer's house. This online diagnosis and education process often provides the exact information the taxpayer was looking for, meaning they may voluntarily decide not to proceed with their appeal, saving time and resources on both sides. If they do proceed with the filing, the system prompts them to provide all the relevant materials automatically, meaning no agency staff member has to pursue open case files to request additional information. The taxpayer can save their submission and return to it at a later date, when they have all the required documentation ready. Once submitted, the appeal is electronically signed and time stamped, so both the Assessor and the filer have confirmation of submission. The submitted information can also be downloaded as a printable PDF file that looks identical to the paper form that the filer would have submitted if they had gone down to the Assessor's Office in person to file the appeal.

Once the filing is complete, the appeal automatically appears in an advanced case management system. The taxpayer may only have one appeal in the system, so their interface is very simple and uncluttered, showing the most updated information on their one case. An agent filing on behalf of many taxpayers may require an interface with more complexity, enabling case sorting and batch actions. The Assessor's Office may have thousands or tens of thousands of cases at one time, so they require very advanced tracking and organization tools. The interface must adjust dynamically based on the user role and the case volume. The system may automatically assign appeals to different agency staff based on geography or appeal type, who are then automatically notified of the new filing. Case administrators within the Assessor's Office can also manually route cases to individual agents or adjusters at any time, and update cases statuses.

All of the actions on the case are tracked in a unified workflow, which can be reviewed in detail at any time. Automatic status notifications are sent to the filer every time an action happens on their case, so the taxpayer is always informed about what is going on, without needing to call in to speak to a live person within the Assessor's Office. Every user in the system has their own login credentials, so data access rights can be managed for each individual using the system. No one sees any cases they do not have the access rights to see, which minimizes confusion.

Every appeal has its own virtual "room" that captures all the case details, serves as a document repository, and shows the current state. Both the agency employees and the taxpayer can post messages or questions within their virtual room, which enables an ongoing, asynchronous conversation between the participants that does not require any meeting scheduling or coordination. Agency staff can also communicate with each other about the case in a private discussion thread that is not viewable by the taxpayer or agent who filed the case, which can help with internal knowledge management.

The system is not only intended for cases filed online, however. If a taxpayer submits the previously described postcard that just says, "I think my tax bill is too high," the Assessor's Office can manually enter the case into the online system and work it in the same manner as a case that was filed online. Even if the taxpayer has not provided a phone number or an email address, the system will generate all of the appropriate forms and letters as printable PDFs, on the Assessor's Office letterhead, so the document can just be printed and mailed back to the filer.

The process moves much more quickly because there are no delays associated with information sharing. If a document is uploaded, all of the participants in the room are immediately notified, and they can download it instantly with a single click. If there is a submission deadline, the platform can automatically notify the taxpayer a week in advance, a day in advance, and even an hour before the deadline approaches, all without requiring any attention from employees within the Assessor's Office. Once a decision is rendered, the taxpayer is informed immediately. The efficiency of the process drives much higher user satisfaction, both inside and outside the Assessor's Office.

If a case is escalated from the Assessor's Office to the statewide Board of Appeal or Board of Equalization, there is no need to go through an additional filing process. The participants from the next level of escalation can just be added into the online room as new users, and the state of the case can be updated appropriately so that the interface is appropriate to the hearing/evaluation process. There is no need to move a paper file across the state or re-upload documents into a new system. Once the statewide office is online with this interactive, technology-based process, any county in the state can escalate a case with the click of a button. This feature can eliminate months of delays in the current offline process, which again improves internal and external user satisfaction.

The system can also easily be integrated into other platforms, such as the CAMA systems used by Assessors or case management systems utilized by agents filing on behalf of others. Data is available via web interfaces for individual taxpayers, but it is also accessible through application programming interfaces (APIs), which allow software platforms to exchange data automatically over the internet. Once a case is initiated, the cloud-based system extracts the relevant details about the property from the Assessor's CAMA system and includes it in the appeal. Once a decision is rendered, the system packages up the outcome and delivers it back to the CAMA system so that the records are fully updated.

Another benefit of the system is reporting. Performance of the Assessor's Office in handling appeals is often very difficult to monitor when the appeals are handled offline. Once an online platform is put in place, all statistics on the performance of the office are reviewable in real time. Case administrators can monitor how many appeals are being filed on a weekly, daily, or hourly basis. Average times to resolution can be calculated instantly, and performance can even be tracked on an agent-by-agent basis as well. Also, many Assessors Offices are subject to transparency laws that require case data be publicly available. These online systems can share statistics with the public in real time, and provide appropriately summarized outcome data to meet public reporting obligations.

## HOW COURT CIVIL CASES WORK

When many citizens think of the courts, they first imagine the criminal system. For generations dramatic television shows and movies have portrayed dramatic court scenes of victims and perpetrators facing justice for their crimes. But in fact, most cases resolved in the courts are civil cases. As FindLaw explains the difference, "Civil cases usually involve private disputes between persons or organizations. Criminal cases involve an action that is considered to be harmful to society as a whole (hence, these are considered offenses against the 'state' or the jurisdiction of the prosecution)." There are some areas that overlap (e.g. traffic, where there are some civil infractions and some criminal infractions) but most of the time civil cases involve a dispute between two private citizens over an agreement and there is no harm to wider society.

Civil cases begin when an individual (called the complainant or plaintiff) files a case claiming that another individual (called the respondent or defendant) has not met their legal obligation in an agreement between the two parties. The complainant files the case in order to compel the respondent to meet their obligation or provide compensation. An example would be a contractor paid to add a room onto a house, but who took the money without finishing the work. In situations like this, there is a dispute between two private parties (they may be individuals or corporations) where one of the parties is asking the court to intervene to settle the dispute.

Traditionally complainants have filed their cases with the court in person, but increasingly courts are utilizing eFiling systems to enable parties to file their cases electronically. Once the filing is received from the complainant, the defendant is contacted and asked to file their response. Once both filings are received, the court will assign a hearing date, usually a few months out from the original filing. The parties will convene at the hearing date, make their case to the judge, and the judge will issue a decision (sometimes called an order) indicating the outcome the judge determines to be fair in the matter at hand. If the parties do not comply with the order then there may need to be additional steps taken in order to enforce the outcome (for example, in some debt cases the debtor may have their future wages docked by their employer to pay a debt the court has determined they must repay).

The civil court system has worked for years in this manner, but lately the system has come under increasing strain. A few of the factors leading to this increased pressure include:

**Budget Cuts**: State legislatures around the county have been steadily cutting court budgets, which means that administrative officials within the courts have eliminated staff, which has reduced efficiency. Some cases now have to wait for months and even years before they come in front of a judge. These delays create frustration amongst parties because the dispute remains unresolved for long periods of time, perhaps increasing harms and escalating anger between the involved individuals.

**Self-Represented Litigants**: Many parties cannot afford to pay lawyers to represent then in the courts, so they elect to self-represent. Sometimes referred to as *Pro Se* litigants, these parties are not educated in the law, and they usually get much worse outcomes than parties who do pay to have lawyers help them pursue their cases. The law can be procedurally complex, often with obligations that are not obvious to non-experienced participants. Legal language is also confusing for individuals who have not received legal training. *Pro Se* litigants will sometimes show up at the courthouse or in legal libraries asking for advice about what steps they should take next, and court administrators and librarians are reluctant to help out because they do not want to violate professional rules that forbid them from providing legal advice.

**Changing Citizen Expectations:** Much like in the area of Property Tax Appeals, Court users have the experience of using interactive technology in other areas of their lives, and they bring those expectations to the courts. When court processes require parties to file documents or show up in person at the courthouse, or to mail or fax forms or evidence, the process seems inefficient and out of date. Younger individuals in particular are extremely reluctant to participate in a process that only takes place offline and face-to-face. This creates a high barrier to entry that many potential litigants are unwilling to climb over, which encourages them to pursue redress through other channels.

## Case Study #2: Filing a Small Claims Case – Paper Process

In one of thousands of county court houses across the United States, litigants sit in court rooms waiting for hours to have their case heard by a judge or magistrate. Ronald, a small businessperson, claims that he has not been paid for his work cleaning the carpets at a local record store, but he does not have a signed agreement indicating that the record store owner agreed to Ronald's price estimate. He'll have to return to his office to find the paper work and come back another day. Ronald doesn't know if he'll be able to return – taking a day off work to come the courthouse has already cost him money he could have been making working for other customers. He may just have to give in and lose the money, even though he did clean the carpets and agreed to the pricing with the record store owner with a handshake.

## BRINGING CIVIL CASES INTO THE CLOUD

Interactive online technology can go a long way toward improving access to the courts, streamlining resolution and improving visibility throughout the resolution process. Many courts have started to deploy ODR (online dispute resolution) systems for civil cases, enabling the parties to work out a resolution via mutual acceptance before the case comes before a judge in a formal hearing.

In a common design for these systems, the complainant still eFiles the case into the court, and the respondent provides their answers to the issues raised. But as soon as that process is complete, the court's Case Management system triggers the creation of an online collaborative workspace (sometimes called a *Resolution Room*) customized for the specific dispute types in question. The complainant then receives an invitation over email or text message to log into the Resolution Room to try to resolve the matter at hand. Once the complainant clicks the link in the email they are taken through a "diagnosis wizard" to get details about the case and some possible forms for resolution. This wizard asks different questions based on the type of dispute that was reported:

**Small Claims Dispute**: If the case involves a low value monetary disagreement (for instance, a debt or consumer protection matter) the wizard collects the details of the dispute (including supporting evidence) and then asks how much money the complainant would like to receive from the respondent in order to fully close the case.

**Landlord Tenant Dispute**: If the case is between a landlord and a tenant, details are collected about the original lease agreement, outstanding payments, and security deposits (as well as supporting evidence). The platform can then help to determine whether the tenant will stay in the property in question or whether the landlord-tenant relationship will end with some level of monetary compensation.

**Family Dispute**: If the case involves a divorce or separation, the wizard can walk the parties through the creation of a co-parenting plan that covers areas like custody, vacation exchanges, child support, and visitation. The co-parents can select their preferred arrangements from pre-configured menus and then negotiate and refine options before jointly committing to the parenting plan.

If the parties achieve agreement on all the areas in question, the ODR system can automatically draft the full text of an agreement and eFile it back into the court case, where it can be certified by the court and the hearing can be cancelled. But even if the parties only achieve a partial agreement the platform can still memorialize the parts they have agreed on, which can make the in-person hearing much more efficient and effective.

ODR systems such as the ones described above have achieved resolutions in more than 75% of incoming civil cases. Also, 85% of resolutions through ODR are achieved during times when the court is not open for business (i.e. late on weekdays or over the weekends). Parties to these cases are satisfied because they get resolutions more quickly, they don't have to take time off of work, and they get their outcomes formally certified by the court. Judges and court administrators like these systems because they enable earlier resolutions, reduce the time required in each case to reach satisfactory outcomes, and move all the court Key Performance Indicators (KPIs) in the right direction: faster case closures, lower cost-per-resolution, higher staff efficiency, and less wasted time.

## CHANGING THE RELATIONSHIP BETWEEN GOVERNMENT AND CITIZENS

The initial compelling rationale for investing in an online case management and resolution system is usually efficiency. Public officials are always looking for ways to make their organizations run more effectively and at lower cost, and these types of systems clearly deliver on those two objectives. But over the longer term, the real impact of these systems lies in fundamentally changing the relationship between citizens and the government. This relationship change can come through empowering individuals with technology tools.

The first type of empowerment is access to information. Right now, there is a very large information asymmetry between government officials and citizens. Given their druthers, most public officials would love for the public to be more informed about how government (i.e. Assessors Offices and Civil Courts) operates. But citizens are too busy to educate themselves about the nuances of their government operations. However, a well-designed, interactive online portal can go a long way to educating the public about exactly how these systems work. Information that may currently be available only inside an agency (such as comparable property valuations, or resolution rates for landlord-tenant disputes) can suddenly be shared in such a manner that individuals can quickly make sense of it. And instead of having to pay for one-by-one education via phone (which usually only happens when a citizen is upset enough to call in) a public official can provide information earlier in the process, when an individual is just trying to learn more and is still shaping their expectations. This kind of early issue diagnosis and resolution is more satisfying for the individual, and much more efficient for agency staff.

The second type of empowerment is access. It is true that any citizen has the legal right to file an appeal or a small claims case. But if filing requires a drive downtown to a government office, for many people, the process stops right there. Maybe it's too far to drive, or maybe their ability to make the trip is impaired in some way due to physical challenges. Maybe they can't spare the time to get down to the office. Maybe they feel they will need professional advice to effectively proceed, and they are worried they won't be able to afford it. Or maybe the length of the process is daunting, and they aren't willing to sign up for a journey that can take months or years.

Some cynics suggest that these impediments to filing are intentional on the part of government officials, who are providing an onerous process in order to discourage people from getting access to redress. In our experience, that is completely untrue. Assessors and judges are public servants, and as such, they want any citizen who has an issue to be able to file a case and get a timely, fair resolution. In fact, many in government are elected officials, so they have a huge incentive to delight their taxpayers so they can keep their job. Government employees are often hamstrung between a longer, administratively complex process that meets all the procedural requirements laid out in legislation, and the frustrations of citizens who want a more streamlined, accessible process. They would gladly opt for a new process that would deliver a streamlined process that met all the required process steps.

A third area of empowerment is visibility. Filing an case currently can feel like putting a note in a bottle and tossing it into the sea. Filers can feel anxious when they have no idea of what's going on. And even when a resolution is finally achieved, the individual may not have any idea as to whether the outcome is in line with outcomes offered to other taxpayers. Managers within government are excited about the better reporting and data visibility that can come from online systems, because it gives them real time intelligence on the effectiveness and efficiency of their operation. But citizens can also benefit from this data visibility, because it enables them to see inside the black box, which demystifies the process and makes them aware of exactly what is going on and who they are working with. It can also contextualize their experience by offering information on the experience of other citizens who have gone through the same process. This transparency can build trust and work to ensure fairer outcomes across the board moving forward.

## MONITORING AND ASSESSING EFFECTIVENESS

Once online systems like this are launched, determining effectiveness on an ongoing basis is essential to realizing the system's full potential. These are learning systems, so continuous improvement is required to maximize the benefits. As problems or points of confusion are identified, systems can be adapted to proactively address them, which generates even more benefits. It is almost impossible to deliver an optimized process flow "out of the box" without some adjustments and tweaks over time. Table 1 lists some objectives of public officials in launching a cloud based appeals management system, and the questions that can determine how successful the new system is in achieving the stated goal.

*Table 1. Measuring the success of online case management systems*

| Goal | Example Questions |
|---|---|
| 1. Improve public relations. | A. What was the public response to the implementation of the ODR system?<br>B. What was the tenor of the media coverage?<br>C. How did communications with citizens about appeals change once the system came live? |
| 2. Reduce costs associated with appeals. | A. Were costs reduced? Which costs?<br>B. Was there a reduction in required support staff?<br>C. Were costs reduced for filing agents or taxpayers? |
| 3. Educate taxpayers about the appeals process so that there can be a reduction in appeals that are filed due to lack of information and to increase the quality of appeals that are filed. | A. Was the quality of resolutions improved (more complete, better data?)<br>B. Was the number of filed cases reduced?<br>C. How many filers began the process, but did not proceed once their issue was diagnosed online? |
| 4. Improve communication channels between the assessing jurisdiction and the filers. | A. Did staff feel like they could effectively communicate with taxpayers?<br>B. Did taxpayers report feeling like they had better access to the assessing jurisdiction?<br>C. Were taxpayers more informed about how their tax bill was calculated? |
| 5. Decrease paperwork and the associated inefficiencies associated with having to move paperwork from staff to staff, department to department, or organization to organization. | A. Was the amount of paper decreased associated with processing appeals? By how much?<br>B. How much time savings was achieved through reductions in paperwork? (Savings per case can be multiplied against total cases to determine total time saved.) |
| 6. Utilize technology to discover dispute patterns—to help prevent and/or more effectively address disputes. | A. Were any patterns discovered? What insights did those patterns provide into appeals cycles?<br>B. Can any of these patterns be implemented in code to auto-decide or score cases that match a particular pattern? |

Once a system is launched, ongoing feedback surveys can determine changes in citizen satisfaction with the new service. Industry standard satisfaction metrics like Net Promoter Score (NPS) can quickly gauge improvements, sorting respondents into “promoters,” “passives,” and “detractors” with just a single question: “how likely would you be to recommend this process to family or friends? (1-10)” (The Net Promoter Community, n.d., para. 2). These data can be collected on an ongoing basis, with results calculated in real time, to provide immediate feedback on process changes and improvements. Other assessment tools such as Curhan, Elfenbein, and Zu’s *Subjective Value Inventory* can determine citizen satisfaction with the negotiation process between the citizen and agency (Subjective Value Inventory, n.d.).

## Case Study #3: Filing an Appeal – Online Process

Sheila, a taxpayer, makes her way to her computer system to file an appeal on her property taxes. She does this after she gets off of her shift at 8 PM. The government offices have been closed for hours and the bus system isn't running this late anyway. She knows that her house has been valued too high, but doesn't really know what documents she needs to provide to make a strong appeal. She clicks on the "Appeal Now" button and realizes that all of the information she needs is provided to her through the problem diagnosis wizard. She just needs to scan in and provide her independent appraisal and specific details about the houses that have sold in her neighborhood. She uses the electronic filing form to submit all of her information and documents. After properly submitting, she can see the status of her appeal and what the next steps are. She can check the status of her appeal and negotiate with the Assessor's Office via her smart phone if she needs to. A few days later, Sheila and the Assessor's Office exchange messages. The Assessor's Office staff doesn't have to call Sheila while she is at work, only to get her voicemail. They can have a meaningful negotiation through the internet. The Assessor's Office can even issue a value settlement offer based on the information they received from Sheila. After the exchange of messages and the acceptance of a value settlement offer, the dispute is resolved.

## BEST PRACTICES FOR PUBLIC AGENCIES

While we have focused particularly on property tax assessment appeals and court civil cases in this analysis, it is not difficult to apply these same observations in other areas of public service and public administration. Obviously any agency that collects taxes [e.g. the Internal Revenue Service (IRS)] or license fees [e.g., the U.S. Forest Service (USFS) or U.S. Department of the Interior (DOI)] can implement a similar system to the one described, reaping similar benefits. But such a system could also be beneficial to any agency that is processing cases at volume, such as information requests [e.g. Freedom of Information Act (FOIA)], insurance claims [e.g. the U.S. Department of Veterans Affairs (VA) or Medicare], eligibility hearings [e.g. U.S. Immigration and Naturalization Service (INS), Social Security], whistleblower complaints [e.g. the U.S. Environmental Protection Agency (EPA)], or civil rights violations [e.g. the U.S. Department of Justice (DOJ)]. We have seen similar systems implemented to collect and resolve complaints of wounded soldiers in substandard housing, or to report copyright violations. The private sector resolves tens of millions of cases per year using systems such as these. In looking at the public sphere, we believe the need is even greater. We can see a time when hundreds of millions, if not billions, of public cases are resolved via similar techniques, resulting in much greater efficiency and satisfaction on the part of citizens and agencies alike.

We have distilled the following recommendations for public decision makers looking to leverage the power of online, interactive technologies to support their work of their agencies:

1. Make the leap to open, cloud-based architectures in creating more interactive online interfaces for your constituents. The temptation is always to build incrementally, and there will be some inside your organization who claim that such systems can be constructed internally on top of existing mainframes and older platforms. It's true that it is possible, but in our opinion, it is unadvisable. You will not be able to scale your system effectively if it is based upon an older infrastructure, and your users will complain about the rigidity of your system moving forward.
2. Choose a SaaS platform carefully. Some vendors will claim they have the expertise to put together a nimble system, but instead they are really generating custom software code that you will have the obligation to maintain and update moving forward. Find a partner, not a provider. Make sure the provider you select has extensive experience delivering parallel systems to agencies similar to your own. Ensure that pricing creates a strong incentive for the provider to keep you happy over the long term, not to cut and run once the initial system is delivered.
3. Check out references. Don't believe slick marketing material. Try out live systems yourself, and get honest feedback from other agencies who have worked with the platform for some time. Ideally, speak to end users of the system in addition to the contracting agencies. Make sure you ask about uptime, responsiveness, and the frequency of platform updates and new feature releases. In this case, past performance is an accurate indicator of future success.
4. Look for solutions that are being tested in the private sector. Public innovation is understandably slower than private innovation, and because switching costs are higher in the public sector, providers focused exclusively on the public sector feel less pressure to innovate. An ideal provider will be one that understands the needs of public agencies, and has a robust set of public customers, but who is also providing similar services in the private sector, where the competition and rate of change is more robust. Those providers will be more responsive, and offer more features and frequent upgrades, than those exclusive to the public sector.

## CONCLUSION

It is an oft-cited fact that the Chinese have the same character for "crisis" and "opportunity." Such a dualism is appropriate in the case of government and interactive technology. A crisis is coming for many agencies, who are being subjected to constant pressure to do more with less. The yardstick for success in the future will not be marginal improvements in past performance, but new standards being achieved by technology-empowered organizations in both the private and public sectors. There is little doubt that the pressure to keep up will be intense, and the change such expectations will put on internal operations will be difficult.

But this disruption also represents an exciting opportunity for leaders looking to take their organizations to new heights and to empower their end users. Technology is going to fundamentally alter the traditional relationship between government and citizens, and if the transition is well managed, it will make that relationship more effective and more trusting. In fact, technology holds the promise of more effective partnering between citizens and government, and a more robust and flexible model for democratic governance. There is no reason to fear the empowerment of citizens through technology, so long as public sector leaders are proactive and visionary about engaging and managing it.

Technology is coming, like it or not. Just as in the private sector, those who learn to understand it and leverage its power will reap great benefits, and those who resist it will pay the price. We believe those organizations who open their doors, move into the cloud, and follow best practices already honed in the private sector will end up on the happy side of that equation.

## REFERENCES

Brink, D. (2010). *Web Security in the Cloud: More Secure! Compliant! Less Expensive!* Retrieved from http://www.aberdeen.com/aberdeen-library/6572/RA-web-cloud-security-compliance.aspx

British Columbia. (2011). *Citizen satisfaction with provincial government services, 2010/2011 citizen consultation on service delivery*. Data presented by Darin Thomson of the Ministry of Justice, British Columbia, at the 2012 American Bar Association Dispute Resolution Section Conference, Washington, DC.

British Columbia. (2012). *Civil resolution tribunal act.* Retrieved from http://www.leg.bc.ca/39 th4th/3rd_read/gov44-3.htm

British Columbia. (n.d.). *Introduction to online service delivery*. Retrieved from http://www2.gov.bc.ca/assets/gov/topic/D9B6F4A42E7F6EF9641DDB53AF6D8B10/intro_to_osd/introduction_to_online_service_delivery.pdf

FindLaw Writers and Editors. (n.d.). *Civil Cases vs. Criminal Cases: Key Differences*. Retrieved from https://litigation.findlaw.com/filing-a-lawsuit/civil-cases-vs-criminal-cases-key-differences.html

IAAO. (2010). *Standard on property tax policy*. Retrieved from http://docs.iaao.org/media/standards/Standard_on_Property_Tax_Policy.pdf

IAAO. (2013). *Standard on assessment appeal*. Retrieved from http://www.iaao.org/uploads/Assessment_Appeal_2013.pdf

Pew Research Center's Internet & American Life Project. (2010). *Government online: The internet gives citizens new paths to government services and information*. Retrieved from http://www.pewinternet.org/~/media//Files/Reports/2010/PIP_Government_Online_2010_with_topline.pdf

Pew Research Center's Internet & American Life Project. (n.d.). *Trend data (adults)*. Retrieved from http://pewinternet.org/Trend-Data-(Adults)/Whos-Online.aspx

Rabinovich-Einy, O., & Katsh, E. (2012). Lessons from online dispute resolution for dispute systems design. In M. S. Wahab, E. Katsh, & D. Rainey (Eds.), *Online dispute resolution: Theory and practice* (pp. 39–59). The Hague, Netherlands: Eleven International Publishing.

Rainey, D., & Katsh, E. (2012). ODR and government. In M. S. Wahab, E. Katsh, & D. Rainey (Eds.), *Online dispute resolution: Theory and practice* (pp. 237–249). The Hague, Netherlands: Eleven International Publishing.

Tergesen, A. (2007, November 4). How to reduce your property taxes. *Bloomberg BusinessWeek*. Retrieved from http://www.businessweek.com/stories/2007-11-04/how-to-reduce-your-property-taxes

Tessler, J. (2010). *Poll: Americans connecting with government online*. Retrieved from http://www.pewinternet.org/Media-Mentions/2010/AP-Americans-Connecting-With-Government-Online.aspx

The Net Promoter Community. (n.d.). *The net promoter score and system*. Retrieved from http://www.netpromoter.com/why-net-promoter/know/

The Subjective Value Inventory. (n.d.). *What do people value when they negotiate?* Retrieved from http://subjectivevalue.com/

Tolbert, C. J., Mossberger, K., & McNeal, R. (2008). Institutions, policy innovation, and e-government in the American States. *Public Administration Review*, *68*(3), 549–563. doi:10.1111/j.1540-6210.2008.00890.x

## ADDITIONAL READING

Abdel, W. M. S., Katsh, M. E., & Rainey, D. (2012). *Online dispute resolution: Theory and practice: a treatise on technology and dispute resolution*. The Hague: Eleven International Pub.

Constantino, C. A., & Merchant, C. S. (2006). *Designing conflict management: a guide to creating productive and healthy organizations*. San Francisco: Jossey-Bass.

Getz, C. (2010). *Evaluation of the Distance Mediation Project: Report on Phase II of the Technology-Assisted Family Mediation Project*. Victoria, BC: British Columbia Mediator Roster Society.

Katsh, E., & Rainey, D. (2011). ODR and Government in a Mobile World. In M. Poblet (Ed.), *Mobile Technologies for Conflict Management: Online Dispute Resolution, Governance, Participation* (p. 85). Springer Press. doi:10.1007/978-94-007-1384-0_7

Katsh, E., & Rifkin, J. (2001). *Online Dispute Resolution: Resolving Conflicts in Cyberspace*. San Francisco: Jossey-Bass.

Rule, C. (2002). *Online dispute resolution for business: B2B, e-commerce, consumer, employment, insurance, and other commercial conflicts*. San Francisco: Jossey-Bass.

Smith, S., & Martinez, J. (2009). Analytic Framework for Dispute Systems Design, An. *Harv. Negot. L. Rev.*, *14*, 31.

## KEY TERMS AND DEFINITIONS

**Application Programming Interface (API):** A mechanism for web-based software platforms to exchange data without requiring manual human per-exchange input or initiation.

**Cloud:** A concept for technology design where software and data are hosted remotely on servers located somewhere on the public internet, not locally on an individual's home computer.

**E-Government:** The application of information and communications technologies to the tasks and responsibilities of agencies and institutions within the public sector.

**Mainframe Technology:** A model for technology design popular in the 1970s and 1980s where computing power and data was centralized in a single large platform and accessed remotely.

**Online Dispute Resolution:** The use of technology to help individuals and organizations find resolutions to their disputes.

**Software as a Service (SaaS):** A model of design for technology projects wherein platform functionality is hosted remotely and delivered to end users on demand via a web browser.

**Sunk Cost:** An economic concept describing previously made expenditures that cannot be reversed.

**Uptime:** The time when a software platform is working fully and is accessible to users.

# Chapter 6
# In the [Source] C\ode of the Conquerors:
## On the Need for Culturally-Minded Tribal Law Research Databases

**Nicholas Ravotti**
*Independent Researcher, USA*

## ABSTRACT

*The practice of law requires not only an understanding of the law itself (i.e., what the law "says"), but also the ability to conduct proper legal research to formulate cogent legal arguments in support of one's case. For attorneys practicing before state and federal courts in the United States, this is accomplished through the use of legal research databases that catalog and archive nearly all state and federal trial court and appellate court opinions. For attorneys who practice before the 573 federally-recognized Indian tribal courts, this task is far more complex. This chapter discusses the need for a culturally-appropriate legal research database to bridge the digital divide in tribal courts.*

## INTRODUCTION: WHAT DOES THE LAW SAY?

In the great tradition of sharing knowledge through storytelling, the following experience of a young pre-law student in his first undergraduate law class is presented to introduce the concept of how the common law functions. The professor begins by positing a "simple," if not broad, question to the class: "what does the law say?"

DOI: 10.4018/978-1-7998-3152-5.ch006

The professor picks a bramble bush to demonstrate her point. What does the law say about freedom of speech? Because the class was held in a state within the United States, a student answers "the law says I have the right to freedom of expression."

"How do you know that's what it says," asks the professor?

After a somewhat startled pause, the student gives what is, actually, a far more intelligent answer than the student will get credit for. The student replies "I just know that's what it says from experience."

The professor does not push the issue, but instead begins explaining that the United States adopted the British system of common law upon obtaining its independence from the Crown. The common law is an ever-changing, ever-evolving system of judge-made law whereby judges issue "opinions" which bind lower courts and, by some blend of pixie dust and magic, citizens throughout the land generally respect and follow. "The law 'now' could be very different than it is… 'now,'" explains the professor.

As the semester progresses, the student is assigned a series of cases to read that prove the professor's point: through time, the right to freedom of expression expanded in some areas, narrowed in others, and sure enough, evolved even during the student's short fourteen-week semester. How did the professor prove her point? She relied on shelf upon shelf of archaic, thick, dusty books called court reporters that catalog and archive nearly all the cases decided by state and federal trial and appellate courts throughout the United States. Or at least, that would be how the professor would have made her point until about the 1950s when digital access to legal documents began taking shape. Today, it would be borderline legal malpractice to spend hours in libraries researching the law when the same task could be accomplished in minutes with a digital database.

That point is worth emphasizing—attorneys today who have ready access to digital legal databases can synthesize answers to complex legal questions in minutes, when fifty years ago, it would have taken their colleagues hours or more to do the same work. Technology has fundamentally changed the entire way the law is taught, learned, and practiced. Today, if a lawyer needs to research an area of the law, she can rely on research services such as LexisNexis, Westlaw, or Bloomberg Law, or even Google for fast access to almost any statute or court decision. *Almost*, any. For attorneys and tribal advocates who practice tribal law in the courts of the 573 federally-recognized Indian tribes within the United States, the digital divide can be quite wide. While some tribe's case law and statutes are more easily accessible than others, often times, legal research for practice before tribal courts requires one part good legal research, one part dedication, and one part experience—i.e., as the student from the story above stated, tribal law attorneys must know what a particular tribe's law says from their experience practicing in those tribal courts.

It may shock and surprise many attorneys that knowing the law can mean relying almost exclusively on life experience—it certainly surprised that pre-law student when he eventually stepped out of law school and into tribal courts. But the simple fact is that tribal court opinions can be hard to find. This is partly because very few attorneys will ever appear in tribal courts, so it is not economically feasible for the big legal research databases to catalog these opinions. An unfortunate consequence of so few attorneys practicing in tribal courts is that the legal profession as a whole simply does not understand what tribal courts are, why they are important, or what they mean to Indian tribes or tribal members. When the profession becomes ambivalent, the legal academy follows. Very few law students ever take an Indian law course in law school, and only a handful of state bar exams actually test Indian law as a subject. (Spruhan, P., 2015). Perhaps the greatest example of the legal profession's ambivalence toward Indian law came from the late Justice Antonin Scalia who is quoted as saying "You know, when it comes to Indian law, most of the time we're just making it up." (Youpee-Roll, 2016).

Another reason tribal court opinions are difficult to research is that tribal governments face unique challenges that states and the federal government simply do not. Therefore, some tribal governments understandably do not expend the funds required to funnel their caselaw to the digital databases for cataloging. Combine the fact that the legal profession is ambivalent, and tribal judiciaries may not have the funding to get their opinions into the hands of the legal research databases, and the result is that some tribes' jurisprudence develops and evolves out of sight of the legal profession writ large. For better or for worse, there is a vase unmet legal need in Indian Country, and lo and behold, the legal profession is falling short of meeting it.

One of the unique challenges that tribal governments face is the imposition upon tribal governments of a legal and political structure that may be foreign to some of these tribes' cultures. From many tribes' perspectives, the legal structure and the way the law is practiced within their tribal courts is simply the imposition of the law of the conquerors, and does not necessarily reflect tribal values with regards to teaching, understanding, or working with tribal law. For many lawyers, practicing law successfully in the United States requires formulating legal arguments with one goal in mind: win—at all costs. Does the goal of winning at all costs reflect the cultural values of Indian tribes? With 573 federally recognized tribes, there are 573 answers to that one simple question. To understand the unique challenges tribal governments face, we must first understand at least a small portion of the history of how the United States used its law to bring tribes under its paternal wing, and how those consequences shape the practice of law in tribal courts today.

## A VERY BRIEF PRIMER ON FEDERAL INDIAN LAW

Understanding the history of the United States' policy toward Indians is critical to understanding the current issues (both legal and non-legal) that affect Indian tribal court systems and the tribal members under these courts' jurisdictions. To say that there has been a federal Indian policy is at best an inaccurate understanding of federal Indian law, and at worst a dangerous misrepresentation of the history and current state of affairs in Indian Country. To view tribal court systems through the narrow lens of current Indian policy denies Indians their history, while at the same time depriving non-Indians of a broader understanding of federal Indian law, and more importantly for this chapter, tribal common law.

Because Indian law is not a topic with which many lawyers or legal scholars are well-informed, no discussion concerning tribal judiciaries can proceed without at least a high-level overview of federal Indian law. This section is painfully brief and does not do justice to the unique status of Indians in the United States, nor does it delve into all of the nuances federal courts have applied to Indian law cases. This section does, however, walk through the oscillating positive and negative treatment of Indians in the United States legal system in order to place some of the long-lasting effects of these policies within their broader historical, social, cultural, and economic context, and more importantly, to shed light on how and why tribal court systems today function as they do.

There are three critically important things to understand about Indian tribes and tribal governments. First, Indian tribes have existed since time immemorial. (*Tee-Hit-Ton Indians v. U.S.*, 348 U.S. 272 (1955)). As such, tribal governments predate the existence of the United States government and the several states' governments. It is a common misunderstanding that the United States granted tribal governments their authority and legitimacy. Quite to the contrary, tribal governments *retain* the inherent sovereignty they have always had. That being said, through a series of United States Supreme Court decisions, the independent authority of tribal governments has been brought under the authority of the United States Congress through Congress's plenary authority over Indian affairs derived from Article I, Sec. 7 of the United States Constitution, known as the Indian Commerce Clause.

The second thing to note about Indian tribes is that a tribe is made up of its members who are both tribal members and United States citizens. This is important to note because one's status as a tribal member is not a racial classification, nor is it a citizenship or residency classification. Rather, tribal membership is a political classification. (*Morton v. Mancari*, 417 U.S. 535 (1974)). As such, laws that apply to tribal members are not race-based laws and, in many cases, a tribal member could avoid application of many of these laws by taking the extreme action of disenrolling herself from her tribe.

The final point to note is that the United States Constitution does not apply to Indian tribes. (*Santa Clara Pueblo v. Martinez*, 436 U.S. 49 (1978)). Congress has exercised its plenary authority over Indian affairs to impose upon tribal governments similar constitutional requirements vis-à-vis the tribe's members. These three principals are fleshed out in the brief primer and overview of federal Indian law that follows.

To a large extent, the journey from tribes' inherent sovereignty since time immemorial to present-day federal Indian law is shaped by three decisions from the 1800s colloquially referred to as "The Marshall Trilogy." The Marshall Trilogy established three long-lasting principles with regards to Indian law. First, the Marshall Court established that nations who discover and subsequently colonize land that is *terra nullius* obtain ownership rights to that land. *Terra nullius* is a Latin term which is literally translated as "the land of no one," but in reality means that the land at issue does not belonging to any particular *country*. (Black's Law Dictionary (10th ed. 2014)). The distinction is critical because *terra nullius* does not mean that the land is uninhabited. Rather, land that is *terra nullius* is simply land that is uninhabited by "civilized" people. The obvious consequence here is that European nations, who spent significant amounts of time and resources colonizing the world during the preceding centuries were permitted, under their own legal theories, to lay claim to land even if indigenous peoples inhabited the land prior to colonization.

In the first of the Marshall Trilogy cases, the plaintiffs argued that they held good title to the land at issue since the land was purchased from the Piankeshaw Indians. The defendant argued that even though his title was issued after that of the plaintiffs, he held good title to the land since he obtained it from the federal government. The Marshall Court held that Great Britain obtained ownership title to all colonized land upon discovery and conquest of what is now the United States, and that title to all land colonized by Britain flowed from sovereign Britain to the sovereign United States government following the Revolution. (*Johnson v. M'Intosh*, 21 U.S. (8 Wheat.) 543 (1823)). In essence, *Johnson v. M'Intosh* delegitimized Indian tribal governments and removed their authority to convey tribally controlled lands.

Notably, *Johnson v. M'Intosh* did not take away all Indian rights to their lands. Indians retained the right of occupancy to their traditionally held lands, subject of course to the federal government's right of extinguishment. With this single sentence, Indians were instantaneously placed under the shadow of the United States federal government which could, at any time, rescind treaties, redraw or eliminate reservation boundaries, or even extinguish tribal governments—actions which the United States Congress has undertaken on numerous occasions. This "Congressional plenary power" over Indian tribes continues to impede tribal governments' abilities to exercise the full sovereignty that other governments the world over enjoy.

The second principle established by the Marshall Trilogy is that Indian tribes are independent sovereign states, but only insofar as they are under the guardianship of the United States federal government. (*Cherokee Nation v. Georgia*, 30 U.S. (5 Pet.) 1 (1831)). In *Cherokee Nation v. Georgia*, the Georgia state government sought to obtain Cherokee held lands by extinguishing tribal title to these lands and removing the tribe to lands west of the Mississippi River. Although the Supreme Court granted certiorari to hear the case, it did not rule on the merits because it determined that the Indian tribes are not foreign states as that term is used in the United States Constitution. Rather, the Marshall Court determined that Indian tribes are "domestic dependent nations," which it went on to describe as similar to the relationship between a ward and its guardian.

The decision in *Cherokee Nation* foreclosed the possibility that Indian tribal governments would be dealt with in the same manner as sovereign foreign states. This holding also created what is known as the "federal trust relationship" between the tribes and the United States federal government. The trust relationship has been both positive and negative for Indian tribes. On the one hand, it requires the federal government to act in the tribes' best interests. On the other hand, it secures a dominion over tribes to the federal government. By holding that Indian tribes are domestic dependent nations, Marshall did not give them equal sovereignty as the federal government. However, he also foreclosed the ability of individual states to create Indian policy on a state-by-state basis.

In the final case of the Marshall Trilogy, Georgia, again, attempted to extend its state law authority over Cherokee tribal lands by passing a law that prohibited non-Indians from entering Cherokee territory without a license. (*Worcester v. Georgia*, 31 U.S. (6 Pet.) 515 (1832)). A Christian missionary named Samuel Worcester was already in the Cherokee territory and remained there after the law was passed, until he was arrested and sentenced to hard labor for violating the statute. The Marshall Court held that Indian tribal jurisdiction is exclusive of state jurisdictional powers within tribal boundaries. Furthermore, the Court held that the United States Constitution reserves the right to regulate Indian affairs with the Congress. *Worcester* therefore firmly places the federal government between Indian tribes and states and reserves regulatory authority over tribes in the federal government, exclusive of state jurisdiction.

The synthesis of these three cases is that Indian tribal governments have legitimate jurisdictional authority over tribal matters arising on tribal lands, but that tribal reservations are not above nor co-equal with states. Rather, the relationship between tribal governments and the United States federal government is one of hierarchy: tribal governments are independent, but only in so far as Congress allows them to be. Once Congress chooses to redefine this relationship (either for the better or the worse from the Indians' perspective) then it can exercise its plenary power even against a tribe's wishes. The tribal government can invoke the trust relationship as a shield to undesirable action, but success in this regard requires the Supreme Court to hold that the tribes have a greater interest in the trust relationship than Congress has in its plenary power. For a court that is highly deferential to Congress (and which does not always fully understand Indian issues), this is a stretch, to say the least.

Federal Indian policy between 1850-1890 took the form of removal and segregation. The dynamics of economic expansion due to the Industrial Revolution combined with security concerns pre- and post-Civil War meant that, generally, non-Indians wanted a buffer between Indians to ensure a peaceful coexistence rather than alliances that might threaten the integrity of the republic. Quite simply, states did not want "domestic dependent nations" within their boundaries.

During this time, there was also a movement to "civilize" Indian peoples. Though obviously offensive by modern standards, this viewpoint was largely a pro-Indian and humanitarian stance during the Removal Period. The idea was that Indians suffered from their own savagery, but if they could just be civilized, then they could contribute to society in their own right. Chief Justice Marshall laid out this viewpoint himself in *Johnson v. M'Intosh*, 21 U.S. (8 Wheat.) 543, 589-90 (1823). He stated that

> *[h]umanity, [] acting on public opinion, has established, as a general rule, that the conquered shall not be wantonly oppressed, and that their condition shall remain as eligible as is compatible with the objects of the conquest. Most usually, they are incorporated with the victorious nation, and become subjects or citizens of the government with which they are connected. The new and old members of the society mingle with each other; the distinction between them is gradually lost, and they make one people. Where this incorporation is practicable, humanity demands, and a wise policy requires, that the rights of the conquered to property should remain unimpaired; that the new subjects should be governed as equitably as the old, and that confidence in their security should gradually banish the painful sense of being separated from their ancient connexions, and united by force to strangers. When the conquest is complete, and the conquered inhabitants can be blended with the conquerors, or safely governed as a distinct people, public opinion, which not even the conqueror can disregard, imposes these restraints upon him; and he cannot neglect them without injury to his fame, and hazard to his power.*

Various Christian churches were a driving force behind this movement to remove Indian peoples westward where tribes would be separated and protected from non-Indian hostility, and where government-funded Christian missionaries would live among the Indians to teach them the error of their ways. The benefit of this program was that those tribes who agreed to be removed westward secured their reservations via treaties with the federal government. This was a choice that was, in actuality, no choice at all. Tribes could either give up any claim to their traditional lands and move west to non-traditional lands which would be protected by reservation treaties, or tribes could remain on their traditional lands which would almost inevitably be overtaken by non-Indian settlement or Congressional fiat.

Unfortunately, there were significant differing cultural conceptions between Indians and the non-Indian federal government as to what treaties were and what purpose they served. Broadly speaking, to many Indian tribes treaties were not merely governmental agreements regarding territorial boundaries or reserved rights. Rather, treaties were long-term covenants through which Indians and their treaty partners were obligated to ensure each other's security and welfare. Both Indian tribes and the federal government entered into treaties with the understanding that the treaty would be binding on both parties. However, from the federal government's perspective, treaties were merely an efficient means to secure concessions from Indian tribes for things such as land and resources that the federal government desired. The result of these differing viewpoints is that, when it comes to laws and legal philosophy, tribal court judges today may understandably be concerned that non-Indian perspectives on "what the law says" do not comport with their own perspectives.

As the U.S. population grew and westward migration continued at a steady pace, it became obvious that Indians and non-Indians could not be completely segregated from each other as envisioned during the removal period. In response, Congress passed the General Allotment Act (24 Stat. 388). The General Allotment Act is also known as the Dawes Act, named after Senator Henry Laurens Dawes (R-MA) who was chairman of the Senate Committee on Indian Affairs and who authored and sponsored the bill. The Allotment Act divided reservation land into 160-acre plots that were distributed to the heads of each family, but held in trust, tax-free, by the federal government for a period of twenty-five years. At the end of this period, the title would be conveyed to the family head in fee simple and subject to state taxation. Any remaining reservation land was then parceled and sold to non-Indian settlers. The proceeds from these sales were given to the Indian tribe.

At the end of the twenty-five-year period, many individual Indians lost their allotted land because they could not afford the tax payments, and these parcels conveyed to non-Indian farmers through forced sales under terms and at prices that were financially detrimental to the Indian landowners. Because of these forced sales, many parcels of land within the boundaries of Indian reservations are held in fee simple by non-tribal members. This raises complex jurisdictional questions as to whether the tribe, the state, or the federal government has jurisdiction over certain issues in Indian Country. As such, legal issues in Indian Country are unique in U.S jurisprudence because these are the only cases where a court's jurisdiction depends not only on where the controversy occurred, but also on who the parties are and their relationship to each other—i.e., whether either or both parties are enrolled tribal members or non-Indians

The allotment period ultimately ended with the Indian Reorganization Act (IRA) (P.L. 73-383, 48 Stat. 984, 25 U.S.C. §§ 461-494), which marked a recognition that prior federal Indian policies had failed, and that Indians should be allowed to govern themselves within the boundaries of their own reservations. One positive thing about the IRA is that it was not compulsory upon tribes—tribes could vote to either accept or reject its application to them. Those tribes that voted in favor of IRA application were granted a charter under which their government would be recognized by the Bureau of Indian Affairs. This was an obvious benefit to the historically precarious status of Indian governmental entities. However, the IRA's drawback was that receiving a charter required the tribe to abide by certain stipulations. For instance, many tribes were required to establish some kind of legislative body, such as a tribal council. For some tribes, this was not a drastic change from how they traditionally governed themselves. For others, the idea of a democratically elected council was a cultural anathema.

Following World War II, federal policy toward Indians shifted from that of allowing and encouraging self-government to outright termination of tribal governmental authority and the federal trust relationship. As was the case with previous policies, termination was pursued as an ill-begotten means of alleviating many of the social problems, such as high unemployment, that existed on Indian reservations. Once a tribe was terminated, tribal assets, such as trust land, were typically sold and the proceeds were dispersed to tribal members. Terminating the tribal government also eliminated much of the tribal social capital upon which Indians relied. As such, the status of Indians whose tribes were terminated tended to worsen, rather than improve. Indians became displaced from their geographic locations, which diminished tribal identity and scattered members across the country. At its most basic level, the termination policies of the United States federal government toward Indians demonstrates how what the law says can detrimentally affect an entire people who had no say in what the law says about them.

The termination period ultimately ended in 1968 with the passage of the Indian Civil Rights Act (P.L. 90-284, 82 Stat. 73, (codified as amended in scattered sections of 25 U.S.C.)). Since that time, Congress and the Executive have generally moved toward an era that Indian law scholars call the Self-Determination Period. While the United States Supreme Court has stopped short of explicitly ruling against Congressional Indian policies, it is fair to say that the Court has not been an activist in the area of Indian law. "In the last twenty-nine years, since the start of the Rehnquist Court, the Supreme Court has decided some fifty-five cases in which either Indian tribes were parties or federal Indian law was at issue. The Justices ruled against the interests of Indians in seventy-two percent of those cases" (Wermiel, 2016). The goal of self-deterministic federal Indian policies is generally to afford tribes the autonomy to pursue the social goals they find to be most desirable. There is, of course, a caveat to the present-day federal Indian policy of self-determination: the Indian Civil Rights Act requires that all tribal governments provide similar (though not identical) constitutional protections in tribal courts as those found in the United States Constitution. Because the United States Constitution is not binding on tribal courts, Congress enacted the Indian Civil Rights Act to impose upon tribal courts its own, rather than the tribe's, conceptions of justice.

ICRA does not incorporate all of U.S. constitutional law into tribal courts. Rather, it merely provides a similar framework which tribal governments must abide by, but which tribal courts are free to interpret through their own case law outside of federal constitutional law. For example, tribal members in state or federal courts have the same First Amendment right to freedom of expression as non-Indians in those courts vis-à-vis the United States federal government and state governments. When it comes to tribal governments' authority to infringe upon its members' rights to freedom of expression, however, the tribal government is bound only by the Indian Civil Rights Act, not the First Amendment to the United States Constitution. Under ICRA's freedom of expression clause, tribal governments cannot infringe on tribal members' rights to freedom of expression, but tribal governments are not prevented from establishing a religion. When presented with an ICRA challenge, tribal courts may follow the path already laid by United States courts, but tribal courts could also choose to develop their own body of law interpreting the boundaries of the "Indian Bill of Rights." As such, like many of the federal Indian policies before it, ICRA imposes non-Indian conceptions of the law onto tribal governments. However, ICRA is unique in that it allows tribes some authority to interpret what the law says through their own culture and values.

## PRACTICING LAW BEFORE TRIBAL COURTS

The economics of practicing law are pretty simple: lawyers charge money to clients, the more cases a lawyer wins, the more money that lawyer brings into her practice, and the arena where it all takes place—the courthouse—costs a lot of money for a government to operate. In other words, underscoring each step of the judicial process is money, and a lot of it. As true as this is for the judicial proceedings themselves, it is equally true of the private market resources available to practicing attorneys, including the legal research databases. In order for a lawyer to do her job, she must know what the law says. But what economic incentive do the legal research databases have to serve the unmet legal research needs of Indian Country? In fact, quite a large incentive, at least with regards to one discrete area of Indian law.

Both Westlaw and LexisNexis have Native American research sections on their respective platforms. These sections include a few tribal codes for the 573 federally recognized tribe, and a smattering of tribal caselaw from these tribes. But by and large, these research sections are mostly dedicated to legal treatises of federal Indian law. The reasons for this are simple. One reason is supply-side economics. Tribal judiciaries control the supply of information coming out of their courthouses to the legal research databases. Without an adequate supply of the caselaw or updated tribal codes, the research databases simply do not have the product to put into their databases to sell. With the myriad of pressing issues affecting tribal judiciaries, such as hiring and retaining judges and court staff, it is understandable that reporting court cases to the legal research databases may not be a tribal court's first priority.

The second reason is demand-side economics. As discussed above, because the legal profession is woefully unfocused on tribal law or Indian issues, there is very little demand for tribal jurisprudence in these databases. The one area where demand-side economics results in a high demand for legal information relating to tribes is in the area of gaming. Tribal gaming is a substantial economic driver for Indian tribes, and therefore fits well into the economics of practicing law. As such, you will be hard pressed to find a single Indian law practice that does not involve a gaming practice group. Although demand for gaming law is high, this translates into absolutely no demand for tribal jurisprudence for one simple reason: there is no tribal jurisprudence involving gaming because tribal gaming is completely preempted by the Indian Gaming Regulatory Act (IGRA) (Pub. L. 100-497, 102 Stat. 2467, 25 U.S.C. § 2701-2721). As such, the highest demand for Indian law in the legal research databases is really a demand for federal law as applied to Indians.

It is worth noting that the Indian Reorganization Act and the Indian Civil Rights Act create some demand for tribal law because these statutes are jurisdictional in nature. Tribes subject to the Indian Reorganization Act have tribal constitutions which bind the actions of tribal governments vis-à-vis those subject to tribal court jurisdiction, and the Indian Civil Rights Act allows those challenging the constitutionality of tribal court decisions to seek a writ of habeas corpus in federal district court. To issue writs of habeas corpus, the federal district court must review the full record of the case from the tribal court. Often, this results in the tribal case record appearing in the legal research database as an attachment or related document to the federal district court decision. This has led to an elevated interest in the boundaries of tribal jurisdiction over causes of action, criminal activity, and regulatory authority concerning tribal members and non-members. To the extent that demand-side economics drives the availability of tribal court jurisprudence in the research databases, it is largely focused on these important, though narrow areas of tribal law.

The simple fact of the matter is that the costs of doing business for a lawyer—everything from office space, malpractice insurance, support staff—are all affected by how quickly and attorney can perform the essential tasks of her job. An attorney with ready access to a legal research database can churn through assignments much more quickly, and therefore, much more cost effectively, than an attorney without these tools. The result is self-feeding: the efficient attorney spends less money and gets more clients. The clients receive more efficient and cost-effective legal services and can therefore afford to purchase more legal services to advance and protect their interests. By contrast, the lawyer who does not have ready access to a legal research database is comparatively inefficient vis-à-vis her colleague with the database. She therefore cannot take on as many clients, and the inefficiency drives up the costs of doing business. Those costs are passed onto the client consumers, who must then spend more money to advance or protect their interests.

This simple economic analysis is in no way Earth shattering. It applies to any widget or commodity available for purchase. Efficiency drives down costs, improves the market, and increases value for consumers. So why should we really care if tribal law is not available on a legal research database, especially if there is not a market demand for this information anyway? Quit candidly, we should care because the law is not a commodity to be bought and sold (or at least it shouldn't be). The law only develops and evolves if those who need to go to court can go to court. If the costs of going to court are high, it will deter many from helping to develop the law through seeking redress in court, and going to court is futile unless the litigant and her attorney know what the law says.

For instance, prior to 1973, women had no constitutionally protected right to abort an unwanted pregnancy (*Roe v. Wade*, 410 U.S. 113 (1973)). Prior to 2003, consenting adults could be criminally charged for engaging in sexual activity (*Lawrence v. Texas*, 539 U.S. 558 (2003)), and no state recognized same-sex marriages (*Obergefell v. Hodges*, 135 S. Ct. 2584 (2015)). What do each of these cases have in common? Each case involved private citizens attempting to secure their own rights. The cases built upon each other to make incremental changes. And taken together, the cases expanded the right to privacy under the Fourteenth Amendment to the United States Constitution which was previously not held to protect any of these rights. As a result, all U.S. citizens, whether or not they ever obtain an abortion, engage in sexual activity with a person of the same sex, or seek a same-sex marriage, enjoy broader privacy protections under the Fourteenth Amendment because someone else was able to go to court to vindicate their rights, and their attorneys could craft legally cogent and persuasive arguments based upon what the law says.

In 1973, digital legal research may have been used by both sides' attorneys in *Roe v. Wade*. In 2003, some form of digital legal research was almost certainly used by the attorneys in *Lawrence v. Texas*. By 2015, the digital legal research databases were undoubtedly used by the litigants, as well as the authors of the 149 *amicus* briefs to the United States Supreme Court. Have tribal communities or tribal governments benefitted from the increased availability of legal research resources?

With regards to their rights under the United States Constitution, tribal members' rights are affected in the same way that other citizens' rights are affected by changes in the law. But with regards to the development of tribal law itself, the legal research databases have added little, if anything, to what tribal law says. Unfortunately, with low demand for tribal jurisprudence from the legal profession, and limited supply from tribal courts due to constrained resources, it is unlikely this this will change.

A final point worth noting is that there are currently a handful of large research databases that capture all of the market share for legal research. A market such as this that is monopolized by a few big players but fails to serve a particular need or demographic is ripe for disruption. If there is one thing that has been shown time and time again, it is that Indians and tribal governments are extremely adept at identifying their unmet needs and finding ways to address them. One unique example of this is the steps that the Cook Inlet Tribal Council took in 2014 to preserve and teach its culture to tribal youth. The Tribal Council identified that tribal youth were not receiving the oral tradition of the tribe, partly because tribal youth were spending more and more time in online platforms such as Facebook, YouTube, and video games. Refusing to let its oral tradition fade away, the Tribal Council worked with a video game development company, a writer, and a tribal poet to develop *Never Alone*, a video game that teaches the tribe's culture by following the gameplay of a young Iñupiaq girl named Nuna and her Arctic fox companion. The game was designed to help preserve Iñupiaq oral traditions for modern and future generations who the Tribal Council expects to become increasingly digitally connected.

Rather than waiting for the corporate-driven legal research databases to build a tribal research database, perhaps tribes can design their own models that better reflect their own tribal values. No one more than tribes themselves know the unmet needs in tribal courtrooms. If savvy investors, developers, and tribal governments can work together to build a video game to preserve culture, then the same could be true to develop a legal research database for tribal law that would essentially be a monopoly unto itself. As is usually the case, when tribal interests work together to meet tribal needs, the result is extraordinary.

## CONCLUSION

Those who came to this chapter for advice or assistance with legal research of tribal law have likely turned its pages and reached its conclusion with disappointment, discovering that this chapter is not a how-to manual for conducting tribal law research. What the law says is only one part of the puzzle. The remainder is left to good legal research, wherever it can be done, and on whatever platform or medium available to do it. This is the source code upon which the law is built. As is clear, the source code of the conquerors has failed to meet the needs of tribal court practitioners, and by extension, tribal members who rely upon the development of the law in their respective tribal courts. In this environment, any shift that helps meet and address this need is a positive movement forward for tribes, tribal members, and tribal law.

## REFERENCES

National Jurist. (2009, Nov. 2). *Scare you to death, work you to death, and bore you to death.* Author.

Spruhan, P. (2015, March). Indian Law. *The Federal Lawyer*, 14–15.

Ward, D. (2016). In Her Words: Recognizing and Preventing Abusive Litigation Against Survivors of Domestic Violence. Seattle Journal for Social Justice, 14(2), 11.

Wermiel. (2016). SCOTUS for law students: Indian cases at the Court. *SCOTUSblog*. Retrieved from http://www.scotusblog.com/2016/01/scotus-for-law-students-indian-cases-at-the-court/

Youpee-Roll, A. (2016). *Supreme Court Makes Up Indian Law Decisions*. Retrieved from Indianz - http://www.indianz.com/News/2016/02/18/april-youpeeroll-supreme-court.asp

# Section 3

# Case Studies: Asia

# Chapter 7
# Chinese Citizenry Social Media Pressures and Public Official Responses:
## The Double-Edged Sword of Social Media in China

**James Masterson**
*Morehead State University, USA*

## ABSTRACT

*Widespread use of social media in China is a double edged sword: social media offers opportunities for the government to connect with society, gauge the opinion of citizens in the public domain, and allow citizens to voice their anger when necessary by blowing off steam online rather than in the streets. However, social media also allows citizens to access information outside of China much more rapidly and efficiently and to link up and communicate with other citizens much more quickly. Social media allows users to share texts, photos, and files, making it much more difficult for the government to control information and to thwart organizing for political purposes. In some instances, the use of social media has forced the Chinese government to take actions that it otherwise would not have done or to reverse actions or policies already set in place. The goal of this chapter is to illustrate the double-edged sword that social media poses to government officials in China, particularly high-level party officials in Beijing.*

DOI: 10.4018/978-1-7998-3152-5.ch007

## INTRODUCTION

Zhang argues that since the 1978 market reforms, the communication landscape in China has changed dramatically, with the CCP wanting to manage this development in order "to stay ahead of the unwanted consequences of reform (Zhang, 2011)." Deng and Moore write extensively on how globalization presents to Chinese leaders a double- edged sword (Deng and Moore, 2004). On one hand, globalization fosters economic development, which has lifted hundreds of millions out of poverty and has supplanted communist ideology as the party's source of legitimacy. On the other hand, globalization exposes China to economic reforms that put pressure on the government to increase the rights of its citizens; an expansion that, one day, could call for the right to elect their own leaders. Such changes thus directly challenge the existing political order. Furthermore, due to globalization, new technologies have become widespread. These technologies have provided the state with new tools for surveillance and control of its citizens. Likewise, technology has given millions of people who live in poverty access to vital information in real time. China is now home to 731 million Internet users (Xu and Albert, 2017). This explosion of Internet use has made it more difficult for the Chinese central government to employ old methods of censorship. Citizens are now able to participate to some degree in the policy formation process ("China Struggles to Tame Microblogging Masses," 2011). Today in China, for instance, there are more than 350 million Weibo microbloggers (Magistad, 2012). Microblogging allows citizens to express dissatisfaction with local policy, while simultaneously sharing pictures and other evidence of a policy's ill effects. However, research by Abbott et al. has found that in China those "most likely to mobilize against the state are the least likely to use the Internet" leading to an initial belief by public officials that it is a relatively safe space for people to interact (Abbott et al., 2013). The media openness during the 2008 Sichuan Earthquake, the first case covered in this chapter, would challenge officials' belief in the Internet as a safe space. As microblogging gains popularity worldwide, recent studies have begun to examined the use of microblogging in response to crisis events and have made recommendations about how to optimize these tools to aid in disaster response (Qu, Huang, Zhang & Zhang., 2011). The goal of this chapter is to illustrate the double-edged sword that social media poses to government officials in China, particularly high level party officials in Beijing.

## THE ARGUMENT

The argument I put forth is that the wide spread use of social media in China is a double edged sword: social media offers opportunities for the government to connect with society, gauge the opinion of citizens in the public domain and allow citizens to voice their anger when necessary by blowing off steam online rather than in the streets (Magistad, 2012); however, social media also allows citizens to access information outside of China much more rapidly and efficiently and to communicate with other citizens much more quickly. Through social media, people share texts, photos and files. "Unlike traditional information sources, social media such as Twitter offer ways to retrieve, produce, and spread timely information during emergency events (Ukkusuri et al., 2014). The speed at which messages are posted after being sent, the ease of duplication of these messages, and the difficulty with censoring messages embedded in photos combine to make it much more difficult for the government to control information and to pre-empt organizing for political purposes.

On one edge of the sword, growth in technologies, particularly as it relates to social media, aids the government in its surveillance efforts. This growth provides opportunities for government officials to participate via these social media to interact with citizens and foster economic development via new social media economic market places.

Aside from controlling information on social media through the use of thousands of internet censors, social media is also a useful surveillance tool of Chinese government leaders and policymakers to gauge public opinion, as there are virtually no other government mechanisms to do this. The use of social media by the government to connect to citizens allows local officials to monitor citizen attitudes and sentiments. Recently, this has become more important for local government officials. In an effort to allow citizens to blow off steam and help hold local government officials accountable, the central government tolerates some politically sensitive blogs and communications to take place (Qiang, 2011). The government tolerates the explosion of nationalist sentiment by citizens in response to rising tensions, leading to sharp increases in nationalism in Chinese cyberspace (Hughes, 2000). In 1999, after wide-spread street protests in the face of the US bombing of the Chinese Embassy in Serbia, the government permitted online airing of grievances through a Bulletin Board Service (BBS) on *People's Daily Online* website (Yang, 2008). More recently, in response to tensions between China and Japan over the disputed Diaoyudao/Senkaku Island, the government allowed citizens to express their outrage via social media in order to increase nationalist sentiment. The result, however, was that this was done at the expense of these protests getting out of control and leading to demonstrations and protests in the streets. Rising nationalist attitudes aid the ruling Chinese Communist Party in making citizens feel they are the party that can best provide for and protect Chinese citizens. While it is unlikely that China will slip back into Mao-era radicalism that gripped the country during the Cultural Revolution, there is still potential for widespread instability throughout China; at least this is a perception perpetuated by the party (Breslin, 2008).

The growth in social media technologies also helps foster connections in the ever expanding market place within China. Economic growth is a major concern of the ruling CCP as its legitimacy to rule is largely tied to economic stability and expansion. Economic disruptions are cause for serious concern for Chinese leaders. A 2008 survey by the China Internet Network Information Center found that nearly 25% of China's Internet users use the Internet to purchase goods and services, and nearly the same amount use the Internet to make payments (CINIC, 2008). Instabilities in the social media networks that are caused by the government undermine its efforts to expand economic development in social media markets.

On the other edge of the sword, these same technologies greatly expand citizens' access to information from sources both inside and outside of China. Furthermore, the explosion of Internet access has also created new openings for political actors (Tilley and Tarrow, 2007). This makes it much more difficult for government officials to cover up or distort the reality of problems citizens face. Citizens can also communicate in real time through the use of blogs and BBS. Forty-seven percent of Internet users in China claim to use some form of social media while 28% of Internet users use BBS services to access information (CINIC, 2012). The blogosphere in China has expanded freedom of expression for Chinese internet users and has made it increasingly difficult for the government to control sensitive information. Some citizens have large followings and have organized petitions and protests against the government (Chung, 2011). And while China employs thousands of human monitors, as well as stringent legal requirements for ISPs to provide their own monitoring of web activity, these monitors simply cannot respond to sensitive posts in time. By the time the monitors read such posts and determine it to be subversive or illegal, there are dozens of other posts that have forwarded on the message. The aforementioned China Internet Network Information Center survey found that 25 percent and 20 percent of China's Internet users blog and participate in online chat rooms, respectively (CINIC, 2012). Furthermore, recent findings by King et al. show that government censoring is largely targeted to prevent mass mobilization efforts, not necessarily to eliminate and restrict certain types of speech geared towards government dissatisfaction (King, Pan, & Roberts, 2012).

However, despite these government efforts, citizens are getting more creative and clever at finding ways to leap over the Great Firewall of China. For example, citizens are using replacement words to organize meeting places. Writers also use photographic images with messages written on them so that the message is not restricted to the character limit of the microblog and cannot be censored for particular words since the message is contained in an image (Magistad, 2012).

Protesting via the use of social media has several advantages compared to protesting in the street. First, many social media outlets have allowed for the user to remain anonymous which increases the incentive for the citizen to take greater risks in their characterization of the problem and their demands to remedy the situation. Protesting via social media also reduces the risk that the protester will get physically harmed by PLA, police officials, or from a potentially ensuing riot. Lastly, protesting via social media allows the citizen to easily participate in multiple causes, drawing links to one another and making appeals for resolutions based on outcomes in other incidents. It is much more difficult due to time and budgetary constraints for the average person to involve themselves in multiple protests on the street throughout distant parts of China.

Despite the advantages, street protests in China have increased significantly as Internet usage has expanded. In 1993, official Chinese statistics document 8,700 "collective incidents" while in 2005 this figure increased ten-fold to 87,000 such incidents (O'Brien and Stern, 2008). The Internet has helped to increase contact among citizens that share similar interests to organize to protest acts of injustice by the rich and powerful, corporate policies that are detrimental to employees and/or the environment and government cover-ups of various incidents for one reason of another. For instance, in May of 2000 students began cyberspace protesting Peking University's policy to cover up the murder of a female student and the barring of student vigils to honor the victim. Through their contacts online, the students organized to protest the university's actions (Yang, 2008). According to Yang, hundreds of students protested and the next day the university held a memorial in honor of the student and gave compensation to the victim's family and promised to increase campus safety.

Simply put, Chinese citizens are finding more and more ways to get over, around, and under the Great Firewall at a time when government censors are reducing control over the system by allowing messages to be sent using different media (ie. through photo and file transfers, not just typed text) and when the government is concentrating its monitoring efforts to subvert organizing activity and not at speech critical of the government. It is in this way, with respect to President Reagan's comment that "The Goliath of totalitarianism will be brought down by the David that is the microchip," David is winning the battle, but has yet to win the war.

## SOCIAL MEDIA AND TRADITIONAL PROTEST

The power of the media resides in the belief by government officials and decision makers that the general public is influenced by the mass media, not in the direct influence of the mass media on the general public (Schudson, 1995). Social media in many instances may not directly influence events in China, but rather intervene by providing new, more effective tools to protest leaders which not only makes the work by protest leaders more successful, but allows for far more actors to become protest leaders. Yang suggests that the fact that authorities in China go to great lengths and expenditure to control the media in China suggests that they too believe in the power that media harnesses (2008). Furthermore, ordinary people believe in the power of the Internet as demonstrated by the rapid growth of users in China.

Protest leadership simply refers to those who initiate group petitions, mass demonstrations and other types of collective action that target political power holders (Li and O'Brien, 2010). Li and O'Brien identify four key functions that protest leaders serve critical to the success of a protest. Protest leaders start by framing the issue for the masses by identifying the source of the grievance with local and/or central government policy thereby more easily enabling people to understand the issues in terms of the protest leader. Additionally, protest leaders are instrumental in mobilizing support for their cause through persuasion (Li and O'Brien, 2010). Protest leaders also are instrumental in creating and carrying out the plan of action against the target.

The spread in the popularity of social media does not eliminate the need for protest leaders; rather, it increases their reach and gives voice to protest leaders who would otherwise not assume those risks in public demonstrations. Social media makes being a protest leader less costly, less risky, and more effective, and aids in the work these leaders seek to carry out. Social media is just one such platform used by protest leaders to give voice to their cause through communication with their followers, which sometimes number in the millions. Adding to that, work by Yan Qu finds that in China the "public seeks out information from all available sources in the immediate post-disaster time period to make sense of the situation, regardless of whether the source is considered authoritative or not (Qu et al., 2009).

Social media communication during crisis events, like those discussed in this chapter, can be categorized into four types of responses: sharing, requesting, collating, and integrating information; opinion dissemination; policy proposal and coordination; and emotional support (Qu et al. 2011). For the purposes of this research we a specifically looking at communication dealing with the first three of these categories since we are interested in understanding the role the social media posts from the citizenry influences policy outcomes. To that end, we examine specific posts about five crisis events in China where by social media users share or request information about the event, opine about the event or the government response to the event, and suggest or demand certain policy action by the government or a reversal of government policy already enacted.

Internet usage in China exploded after 1998 when there were just 2.1 million users. Just two years later usage grew more than tenfold to 23 million users. Currently, there are an estimated 513 million Internet users out of a population of around 1.4 billion people (CINIC, 2012). About 37% of Internet users today access the Internet using mobile phones. This is a 7 point increase from 2008. As more than 60% of mobile users are between the ages of 20 and 40 (CINIC, 2012), mobile usage will certainly become much more commonplace as today's youth grow older and web-access via mobile phone becomes less expensive, easier to use, and has greater capabilities with the creation of new mobile application software. More and more Chinese have the ability to share information, pictures, and opinions from events all around China, using the social media applications integrated into their mobile devices.

Research on Chinese contemporary mass movements has focused on the "down and out." That is to say the focus has largely been on peasants who have been forced to relocate their residence to make way for a new factory or residential development, workers who have been laid off, pensioners whose support has been reduced or eliminated, and tax-burdened farmers. Additionally, those affected by environmental degradation as China rapidly develops are also expressing contention both online and in the streets (O'Brien and Stern, 2008). These social groups often resort to protests in the street in the face of adverse conditions. However, students, which make up the largest portion of China's internet users, have, early on, been active in cyber protests. Examples include the 2000 murder of a female student from Peking University where students protested the university's cover up of the murder and barring of student memorial gatherings and the 1999 protests against the US after the bombing of the Chinese Embassy in Serbia (Yang, 2008). However, it remains difficult to know whether or not there are differences in the uses of social media for purposes of protesting among different social groups. This is due to the vast expansion of cyberspace to now include many hand held mobile devices used by hundreds of millions more Chinese now compared to a decade ago combined with the anonymity allowed by many social media sites and blogs. However, research by Gainous et al. lends some support to the idea that those using the Internet in China are likely to engage in higher levels protest against the government, albeit, online. Gainous et al. find that in East Asia more frequent use of the Internet leads to the development of negative attitudes about government, resulting in less obedience (Gainous, Wagner, & Abbott 2015).

Linking actors is crucial to the success of any mass movement. The internet, and social media for that matter, is a particularly useful tool to create such linkages. As Yang argues, access to the internet is becoming much more common place so that there are greater and greater opportunities for citizens to be linked into a cause and to themselves become citizens participating in an online protest, sharing information and opinions on an important contentious political or social matter (Yang, 2008). These tools are increasingly integrated into other forms of media, audio, video, news bulletin posts, file transfer etc. And lastly, Yang argues that Chinese social media is increasingly linked to the outside world, reducing the efficacy of Chinese state control of the internal media. Social media helps to create links between cyberspace and physical space by organizing protesters, sharing images of the events in real time over the internet, and creating discourse about the events whereby the most popular slogans and sayings get translated into signs and posters at the protest. It is in this way that social media helps to foster these links. It is not that social media alone creates these changes in political outcomes, but rather, is more efficient and powerful at creating links among interested individuals for common action. Cyberactivism can pave the way for political change by becoming a major trigger for street activism; "energizing civil society and encouraging civic engagement; aiding the mobilization and organization of protests and other forms of political expression; and promoting a new form of citizen journalism by which ordinary citizens were able to express themselves and document their own versions of reality" (Khamis & Vaughn, 2011).

## SOCIAL MEDIA IN CHINA

To better understand how social media presents Chinese leaders a double-edged sword, it is useful to briefly discuss the most popular social media applications in China and their capabilities. While it is well-known that popular American social media outlets such as Facebook, Twitter, and Myspace, are blocked in China, the Chinese have essentially developed their own versions of these programs. In fact, some of the most popular social media in China have greater capabilities than the most popular programs in the United States. Perhaps by some measure, the most popular Chinese social media outlet is a program called QQ by the software company Tencent Inc. QQ boasts more than 500 million registered users for its instant chat feature that is cross-platform compatible. QQ International allows English-speaking users to join the mainland China QQ network as well. Computers, cell-phones, and tablets all have Tencent applications available to run QQ. Within the QQ chat, users can create groups allowing large numbers of members to connect and share information. Users can chat via text, audio and video peer-to-peer and can upload files and folders even when the user is offline. The upload is made available to the recipient when the user logs in within fourteen days of the file's upload. In addition to these real-time chat features, members of QQ can create a Q Zone page, similar to a Facebook page. Members can store photos, maintain a Blog, and share messages with their Q Zone friends. Friends can start discussion threads and post comments on the Q Zone page for others to read and respond.

Another very popular social media program among the Chinese is Sina.com's Weibo (China's Twitter). Weibo has taken some functions from Twitter, Groupon, Facebook, and YouTube and assembled a product that is much more functional than Twitter. The service allows for embedded images and videos within its posts. The site has served as a major rallying point for Chinese citizens to share real-time developments during disasters and rapidly developing events. Like QQ, Weibo has a host of monitors that censor its content before it reaches the eyes of the government monitors. In addition to pressures the in-house monitors face from the government, they also face self-censorship pressure from investors. Rabkin (2009) reports that Chinese media outlets experience censorship pressures to protect their financial interests, not solely their legal interests. Corporate investors and licensure agencies have a direct impact on what these media outlets allow to be published.

Jeremy Goldkorn, director of Danwei.com, a Chinese social media firm, recognizes that Weibo can aid organizers and protests against government policy, but that Weibo has also become an asset to the ruling Chinese Communist Party (Olesen, 2012). The government can allow citizens to use the social media as an outlet for the frustrations, keeping them online, instead of in the streets. Furthermore, such public airing of grievances allows the central government to monitor problems at the local level and helps them hold local officials accountable.

Lastly, among China's most popular media outlets is Renren. Unlike the previous two social media programs discussed, Renren users must use their real name. This is a limitation that this particular social media has encountered, as users that prefer anonymity are not likely to use Renren as their primary outlet. As of 2011, Renren had nearly 125 million users. It is very similar in design and functionality to Facebook, however, it is not nearly as competitive in its domestic market as Facebook is in the United States when viewed in terms of users. Domestically, Weibo and QQ have more users than Renren, while Renren struggles to remain profitable as it has recently launched its IPO. For instance, Renren just recently turned a modest $800,000 profit in the second quarter of 2011, its first profitable quarter (Rapoza, 2011).

WeChat is a free application launched by Tencent in China in 2011. It is a mobile media platform having similar functions as QQ with regards to messaging and with functions similar to Weibo allowing users to maintain a site for blogging. Through WeChat users can send texts, pictures, videos and voice messages via the Internet, as well as several service plug-ins such as "Public Platform", "Shake",

"Sweep", "Raft Bottle" and "Friends Circle". It integrates SMS, phone and email, and binds friends to mobile phones. It provides users with a convenient social space and communication platform (Chen, 2013).

WeChat is a complex platform that integrates text, voice and call communications, social networking, information sharing and commercial exchange, a fundamentally new aspect of social media platforms, all within a single application. These additional social and commercial functions make the communication properties of WeChat more complex. It's more than just a self-media. Some scholars refer to this commercialization process as a kind of "de-mediation." In China, WeChat's voice chat function replaced the conversation, allowing users to quickly send a voice message that can be as long a one minute. The WeChat payment function is quickly replacing the need for cash (Yu, 2015). From street food vendors to taxi drivers to mall merchants, nearly every vendor in China accepts WeChat's payment system as a means to exchange money for goods and services.

WeChat's public account service platform has become the main channel for people to obtain news information and express their opinions about local, national and international events. Once the WeChat users follow a public account, they receive information posted on the account. Users do not have to be invited to follow a public account, but self-select which public accounts to follow. WeChat creates a new sharing mode based on the functions of existing social media platforms such as QQ and Instagram. (Yu, 2015).

Though WeChat has become a major force in China's mobile network communication, benefiting its citizens, there are downsides for the government to high levels of citizen use of this platform. WeChat has the potential to cause difficulties in managing communication content and flows by the government for oversight. For instance, WeChat has a complicated user identity, few barriers to entry for new users, and an open voice information dissemination, all of which make it difficult for the Chinese government to fully manage and control the WeChat community. Although the current network monitoring technology in China is relatively mature, it is still difficult to monitor and capture sound files. WeChat plays an important role in the dissemination of information regarding major events and crisis, which have the ability to significantly influence public opinion. Therefore, in order to control public opinion, the Chinese government must often launch "public opinion guidance" directives forcing public accounts to post government messages (Jing, Zhou & Ma, 2014).

These five social media programs and others in China allow users to rapidly disseminate still and video images and audio and text messages to connect in real time and at a later time with one another. WeChat allows users to send money instantly from user to another, and to use WeChat to pay for goods and services. This enables activists to help support each other financially in support of common goals. All of the social media platforms allow users to form groups and followings and allow users some degree of anonymity, depending on the program. Despite heavy censorship by the software company and government monitors, users have found creative and innovative methods for getting around such censorship. In the section that follows, I will illustrate how social media presents a double-edged sword to the leadership in China. Discussed are three events that have drawn significant attention of Chinese citizens. Some of these citizens have sought to use social media to change government policy by rapidly disseminating information detrimental to the course of action the government had initially pursued or they were encouraged by government policy into action by the means of communicating through social media.

China is certainly not unique in attempting to control access to the Internet. Recent events in North Africa during what has been called the Arab Spring have illuminated the control over communications states have at their disposal, but also the difficultly states face in striving for absolute control. In Egypt as millions protested in Tahrir Square for regime change, Hosni Mubarak ordered the state cut off from the world and turned off the internet and cut mobile phone communication networks (Alexander, 2011). However, he was too late. Millions of Egyptians had already organized protests around the country using Facebook to communicate. In this case, instead of just blocking popular social media sites (which was tried days before the Internet blackout), the government shut down ISPs effectively shutting down more than 85% of Egypt's Internet connection to the world (Arthur, 2011).

Similarly in Libya, before his ousting and subsequent death, Muammar al-Qadaffi cut off access to certain Internet sites within the state. However, this had little impact as less than 6% of the population had Internet access. In contrast to state run media reports, Cloudflare CEO Matthew Prince says the shutdown of Syria's Internet access was entirely at the government's hands. "The Syrian Minister of Information is being reported as saying that the government did not disable the Internet, but instead the outage was caused by a cable being cut," writes Matthew Prince, CEO of CloudFlare. "From our investigation, that appears unlikely to be the case" (Gross, 2012). According to Prince, Syria has four separate connections to the world and all four would have to go down for there to be a nation-wide shut down for the Internet grid, which is what happened. Therefore, an explosion near one of the connections could not knock the entire nation offline as the government has purported the cause to be.

In a more extreme case, the North Korean regime restricts the use of the Internet for nearly the entire nation. Those who can access the web are limited to North Korea's own intranet where each user's identity can be closely monitored and tracked. Things are slowly changing in North Korea. AP Journalist Jean Lee reports that foreigners are now allowed to bring in cell phones and access North Korea's 3G network, including popular social media websites such as Twitter. She claims that there are two parallel systems in North Korea, one for residents and one for foreigners (Griggs, 2013).

According to a 2012 report by *The Guardian*, Internet censorship is quite pervasive around the world. However, it is found largely in Asia and the Middle East. Where data exists, no countries in North America, Australia or Western Europe were found to censor the Internet for political or social material. Evidence for Venezuela and Columbia showed selective censorship of social material but found no evidence of censorship of political material. In contrast, evidence suggests that in nearly every major country in Asia, with the exception of S. Korea, Afghanistan, Iraq, and Israel, there is a suspect level of censorship for political or social material (data was not available for Japan). China, Vietnam, Iran and Syria are listed among the few countries that show evidence of pervasive political censorship on the Internet. Interestingly, China's highly populated democratic neighbor shows selective levels of political and social censorship on the Internet (Blight, Rininsland, Rogers and Torpey, 2012). The most significant difference between Internet censorship in India and China is that India works with ISP providers, Internet search engines, and the courts to block content on a case by case basis, whereas China employs nearly 50,000 human monitors to censor web activities inside the PRC and access to websites outside the PRC.

## THE CASE STUDIES

Four cases are introduced in this chapter to illustrate, across different events, how the use of social media influences government policy and action. In the first case, the use of social media during the 2008 Sichuan Earthquake (also referred to as the Wenchuan Earthquake) led to a greater accountability of government officials responsible for aiding the victims of the earthquake and accounting for the death toll and cause of destruction.

The second case study demonstrates the use of social media by the government to allow and even encourage expressions of anger towards the Japanese over the recent island dispute. In this case the government attempted to use social media as a means to achieve a particular end. While it seems the government is using social media to investigate and bring to justice violent protestors and to increase nationalist sentiment for its own self-preserving ends, it can also reasonably be seen that the government is allowing citizens to vent their frustrations with internal Chinese policies associated with how the government is handling its response to the Japanese, so as to lessen the chances of a larger public protest on the streets.

The third case represents a situation where the use of social media puts significant pressures on the government to pursue a different course of action than it otherwise would. While the government did not cause the Wenzhou train wreck, central leaders were publicly boasting of China's competitiveness in the high speed rail industry.

Their creditability rests to some degree with their policies to develop this industry within China. When the wreck occurred citizens were appalled by the government's decision to hastily bury the bodies and the wreckage from the event without a thorough investigation into the causes of the crash. The outrage expressed by citizens carried through various social media in the form of blogs and photos put sufficient pressure on the government to decide to investigate the crash site, an action they otherwise would not have pursued (Magistad, 2012).

Lastly, the final case illustrates a situation where citizens rose up against their local government in direct response to actions or policies pursued by these governing officials. It is different from the preceding two case studies in that the protests were not organized by any government officials (as in the island dispute) in hopes of assuaging the anger of the villagers that had lost property in this land deal, nor was the government trying to cover up devastating results of a culmination of failed policies (as in the Wenzhou train crash). This case is similar to the preceding two in that citizens rose up to try to reverse government policy from what it otherwise would have been. In the village of Wukan, citizens organized protests and eventually ran local leaders out of the village and negotiated a deal with the provincial government to elect new village leaders. Furthermore, the provincial level government had a stake in hearing the citizens' protest in an effort to help identify and weed-out corrupt local officials. These cases show the use of social media to link together individuals in protest of government actions.

## 2008 Sichuan/Wenchuan Earthquake

This case study will discuss the impact that social media had during the 2008 Wenchuan Earthquake to hold government actors accountable for their response to the crisis. As will be shown below, pressures created by citizens using social media as a platform to raise awareness of and demand accountability for the number of school children whose life was lost during the earthquake ultimately led to the government to publicly release official data a year after the earthquake and a local investigation into the shoddy construction techniques used to build many of the schools in the most devastated regions. However, a full investigation into the construction failures of the schools in Sichuan that collapsed during the quake has never been conducted.

May 12, 2008 was a typical afternoon in the foothills of the Longmen Mountain range, where young children at school were napping after their lunch, older children studying in class, while the rest of society carried about their day as any other. At 2:28 pm in the small town of Wenchuan, north north west of Chengdu, a devastating earthquake measuring 7.9 on the Richter scale hit this town and several counties and cities located on the eastern edge of the Qinghai-Tibet Plateau in the western Sichuan Basin (Parsons et al. 2008).

By the end of the two-minute long rumble, the earthquake had damaged more than 4 million buildings in the province, of which at least 6,898 were school buildings ("Sichuan earthquake killed more than 5,000 pupils, says China," 2009 and Wu, 2008). Within two months Chinese officials announced that the Wenchuan earthquake had killed 69,227 people, injured 374,643 and left 17,923 missing. In this earthquake, students suffered heavy casualties as a result of the time and day when it occurred. Many young school children were taking a nap at the time the earthquake stuck, delaying their reaction to the event, time that may have allowed them to get to safer locations. It would take a year and pressure from non-governmental groups for the official tally of the casualties of the school children to be released. In May of 2009 Government officials reported that the earthquake killed 5,335 students while 546 children went missing (China: 5,335 students killed or missing after 2008 quake, 2009).

Social media was one of the first places people turned to for information about the earthquake. Turning to Twitter and YouTube, at the time not banned in China, users posted pictures, videos and comments about the quake within nine minutes after it struck (China Feels the Quake Online, 2008 and Moore, 2008). Research examining social media postings on a Chinese forum called Tianya shows that there were more than 14,000 threads created within the first week of the disaster (Qu, Wu, and Wang, 2009). The Chinese government's swift reaction to disaster relief and the loose restrictions on media coverage, including the openness of social media coverage, of the incident received international recognition (Kaufman, 2008). As such, the Internet was one of the most frequent places visited by users for information about the disaster and to share information and express their opinions (Qu, Wu, and Wang, 2009). "Using the Internet netizens covered the earthquake and salvation, called on people to donate money and material, offered information and suggestions for salvation, supervised the work of government, and so on," according to PhD candidate Xu Liang, wring online for an online Korean publication (Xu, 2008). Many Chinese citizens turned to the use of Google Maps to create a community response grid that shared information about the devastation and allowed public officials to contribute responses (Winerman, 2009). In a study by Winerman on the use of social media during crisis communication, she found that one social media user in China was able to direct a military helicopter, whose pilot was unfamiliar with the area, to an appropriate landing site with a detailed description by posting it to an online forum. Both Winerman and Xu recall this story in their research and report that the information eventually found its way to the Chinese military (PLA), who landed a helicopter in the spot described with aid materials (Winerman, 2009 and and Xu, 2008).

However, after nine days of coverage of the school building collapse from various outlets, negative news regarding China's use of poor construction standards and oversight of the construction projects were reported in the news media and on the internet by local people, journalists and volunteers, who often referred to these as "tofu projects" (Levin, 2013). Evidence supporting these claims began to grow quickly in the days after the tremor hit. Structural engineers and quake experts, at the request of the *NY Times*, analyzed photos of the wreckage and found that the use of poorly reinforced steel, hollow slab floors and too little cement was found to contribute to the collapse of many schools (Yardley, 2008). An NPR report shortly after the quake documented shoddy construction in the town of Beichuan, where about 21,000 people perished in the quake. The report cites poor construction practices as a main cause of the collapse of buildings where many of the victims died (Five years after a quake, Chinese cite shoddy reconstruction, 2013). Within weeks of the event, a series of reports and investigations detailed the poor construction of schools buildings in the affect area (Bandurski, 2009). The families of the injured students strongly criticized the government's involvement in these practices, believing that local officials and builders had cut corners in school construction. As evidence of corruption and corners cut during construction mounted, many cited the fact that other nearby buildings had hardly been damaged after the earthquake (Trindle, 2008). And in some areas, schools were the only buildings that collapsed, increasing people's suspicion of improper supervision and corruption of government personnel during the construction process of the schools (Police break up protest by parents of China earthquake victims, 2008).

As noted above, government officials were reluctant to release official data on child casualties from the earthquake. Despite initial openness of the media to report on the earthquake, the government began restricting any negative reporting about the relief effort and links to the vast devastation caused by the tremor, fearing it could lead to wide spread social unrest. The government began recording the number of dead and missing within hours of the quake, however, the authorities refused to release the figures at the time (Sichuan earthquake killed more than 5,000 pupils, says China, 2009). Requests for official accounts of the child casualties went unanswered (Ai, 2018). According to Geoffrey York of the *Globe and Mail*, in some cases, "parents who tried to file lawsuits were dragged away from the court building by police." The *Guardian* also reported incidences of protesting parents being "detained or warned against speaking out" (Sichuan earthquake killed more than 5,000 pupils, says China, 2009). According to an Amnesty International report, *Justice Denied*, threats faced by parents of the victims were so bad that the international advocacy group "called on the Chinese government to stop intimidating parents and relatives of the child victims, who face harassment and arrest as they seek justice for the dead and injured" (Coonan, 2009 and Amnesty International, 2009).

York found that in an attempt to quiet media attention to and online discussion of the construction of the collapsed school buildings, authorities "managed to suppress some of the debate by sending police to seal off the collapsed schools and prevent memorials on the sites. Beijing has ordered the media to focus on uplifting stories about earthquake heroes and reconstruction efforts, rather than the poorly built schools (York, 2016). A UNICEF report published two years after the disaster cites that "more than 12,000 schools in Sichuan and 6,500 schools in Gansu were damaged or destroyed. According to York, near many of the collapsed schools were "surrounding buildings [that] often survived intact."

In an attempt to uncover the number of child casualties to the earthquake, separate campaigns were started using social media to dissemination information about the devastation, share opinions with users about links to the destruction, such as poor building construction of school buildings, and recruit volunteers who would visit the devastated areas interviewing families to account for missing and dead children (Ai, 2018, Bristow, 2009, Ford, 2010, and Wong, 2010)

According to reports, Tan Zuoren used social media to organize an online campaign to independently investigate the number of schoolchildren killed in the earthquake and the reasons the school buildings collapsed (Ford, 2010). He created a 5.12 Student Archive to account for the child victims of the quake in a database by asking people who lost their children to contribute the details to his victim database (Kwong, 2011). Bandurski found that Tan "made scores of trips into the disaster zone, and an incomplete citizen survey of 64 schools that he released confirmed the details for at least 5,761 students that had died in the quake, the vast majority due to shoddy school buildings" (Bandurski, 2009).

Internationally known artist, Ai Weiwei turned to social media to recruit volunteers to visit the devastated areas to interview with parents of the dead schoolchildren to develop a list of them in an effort to account for all of the deaths of the children. He eventually published his own list of the children killed (Wong, 2009).

A Sichuan school teacher, Liu Shaokun, traveled throughout the devastated areas taking photographs of collapsed school buildings and posted those online. According to the advocacy group, Human Rights in China, Liu expressed anger at the poorly constructed buildings during media interviews (Sichuan Teacher, Liu Shaokun, was released to serve his reeducation-through-labor sentence outside of labor camp, 2008).

For his part in pressuring governmental officials to publically account for the dead schoolchildren and investigate the factors leading to the collapse of school buildings, Tan Zuoren was put on trial and detained, a fate shared by many other activists and protesters calling for greater accountability of the local government with regards to why the school buildings collapsed when so many buildings near them did not and to account for the dead children (Coonan, 2009). He was sentenced to five years in prison, though according to Peter Ford, Tan was sentenced for "incitement to subversion" and not specifically for his activist work in investigating the collapse of the school buildings nor compiling a database of the dead school children who were victims of the earthquake (Ford, 2010). According to Ford, Tan's lawyers said this was the heaviest sentence the court could impose. Despite this, Tan's work continued after his release where he and 100 volunteers wrote to state leaders to appeal for an official investigation into the substandard school building construction (Lao, 2018).

As a result of Ai Weiwei's work to draw attention to the student casualties of the quake he claims he was beaten by police in 2009 in preparation of his testimony at an activist's trial (Lai, 2012). Several of Ai Weiwei's volunteers were detained by the police. "The police confiscate your materials and give no receipt. They send you back to Chengdu by car, and tell you, 'Don't let us see you again. You have broken the law" (Kuhn, 2009).

Liu Shaokun was detained by police on June 25$^{th}$ for his role in disseminating pictures of collapsed school buildings. He was officially charged with "disseminating rumours and destroying social order" according to the *Telegraph* (Chinese man held over earthquake photos, 2008). He was sentenced to one year of reeducation through labor, though he was released early to serve the remainder of his sentence outside of the labor camp (Sichuan Teacher, Liu Shaokun, was Released to Serve his Reeducation-Through-Labor Sentence Outside of Labor Camp, 2008).

Nearly a year to the day, Chinese officials relented and finally publicly released the count of schoolchildren that died in the earthquake. They put the figure at 5,335 schoolchildren or students that died or remained missing as a result of the quake. Though the government began counting the dead and missing within hours of the disaster, it would take the outcry of citizens protesting locally and using social media to reach a larger audience to demand the release of these figures. Ai Weiwei said that the "announcement was a sign that the government may be caving in to the 'pressure of the common people, pressure from the media. (Sichuan earthquake killed more than 5,000 pupils, says China, 2009).

Nearly a month after the quake, officials in Shifeng promised an investigation into ten schools that collapsed (Buckley, 2008). In April of 2009 Sichuan officials were publicly declaring that there will no investigations into the construction of the school buildings, despite Beijing's initial pledge that there would be "in-depth, speedy investigations" carried out (Sichuan earthquake: government prohibits investigations into collapsed schools, 2009). After more than eleven years of online and on-ground protests an official investigation has never been carried out (Lao, 2018).

In 2008, social media was in its nascent stages of development in China. Few Chinese platforms existed and few Western platforms were banned. Government officials had little understanding of the impact that social media could play in policy outcomes and policy exposure. By the time the next major earthquake hit in Sichuan in the town of Lushan, both the citizens and public officials would have a better understanding of how to harness the power of social media to aid in the relief effort. Government officials and volunteers gained valuable lessons from the 2008 disaster in Wenchuan to more effectively direct their rescue efforts in Lushan. Volunteers efforts to organize and outsmart the government when trying to keep them away "were enhanced by Sina Weibo" which did not exist during the Wenchuan crisis and had more than 500 million by the time the Lushan quake hit (Levin, 2013). Many citizen aid efforts were directed online to recruit aid materials and volunteers, sometimes at odds with the government. The "rise in online activism has forced the governmeny to adapt" (Levin, 2013).

## Diaoyudao/Senkaku Island Dispute

The Diaoyudao/Senkaku Islands are located in the East China Sea, approximately 120 nautical miles east of Fuzhou, China and ninety nautical miles north of Japan's Ryukyu Islands (Manjiao, 2011). They are comprised of five uninhabited islets and three barren rocks, totaling a land area of less than seven square kilometers (Manjiao, 2011). Japan seized the islands from China as spoils of war in 1895. However, following the issuance of the Cairo Declaration, Japan was required to return "all lands taken from China during the 1894-95 dispute and afterwards (Diaz, Dubner, and Parent, 2007). However, the islands are still formally administered by Japan.

The dispute over the islands has been ongoing since the early 1970s, when rich oil resources in the seabed were discovered, prompting both China and Japan to claim sovereignty to the islands. These islands also serve as an important delineation boundary between China and Japan. This is important for claiming Exclusive Economic Zone (EEZ) privileges, which would expand territorial waters stretching 200 nautical miles from the coast. EEZ privileges are also necessary to obtain the rights to any natural resources such as oil. Over the years, the Islands "have become closely intertwined with the national identity of the two countries," prompting nationalist sentiments anytime a showdown over the Islands occurs (Manjiao, 2011).

In September 2010, tensions rose following the two-week detainment of a Chinese fishing trawler and its captain. The trawler had deliberately collided with Japanese inspection vessels in the waters off the Islands (East Asia Territorial Disputes: Geopolitical Implications, 2012). According to the Korea Herald, "Japanese authorities wanted to establish a precedent by prosecuting the skipper of the Chinese fishing boat under its domestic criminal procedures" ("China-Japan Dispute," 2010). Beijing responded by suspending ministerial level contact, having travel agents cancel tour programs to Japan, and halting all Japanese-bound shipments of rare earth metals essential for high-tech manufacturing ("China-Japan Dispute" 2010). After little more than two weeks, Japanese prosecutors released the 41-year-old Chinese skipper, citing the implications on the future of Japan-China relations. Tokyo denies giving input to the prosecutors; however this statement is widely dismissed as false.

Another flair up related to the island dispute occurred again in April 2012, when Tokyo's prefectural governor, Shintaro Ishihara, proposed that the Tokyo Metropolitan Government purchase the islands from private Japanese owners. Hoping to diffuse the situation, the Japanese government made a counter-proposal in July 2012 that it would purchase the islands instead of Ishihara, who is Japan's most well-known nationalist politician ("East Asia Territorial Disputes: Geopolitical Implications," 2012). However, China warned that any attempt by the Japanese to purchase the Islands would not change its claim over the islands.

In the PRC, protests against the Japanese turned violent as a result of the central government of Japan purchasing three of the islands, resulting in damage to Japanese-brand cars, dealerships, department stores and businesses (Zhai, 2012). Some Japanese-funded companies closed. In Shanghai, reports say an angry man drove his Honda Civic to a dealership and set the vehicle on fire. Photos of the burning car, in front of a banner reading, "Don't forget our national shame," were circulated on Weibo ("In China, anti-Japan protests over disputed islands get some zing," 2012). A video clip released on September 26 shows the brutal beating of a Toyota car owner. The video went viral on Chinese video-sharing websites and microblogging services; many citizens condemned the video and called instead for "rational" patriotic actions. In the video, Li Jianli tried to protect his car from protestors in the northwestern city of Xi'an. A protestor attacked Li, hitting him several times in the head using a steel lock, leaving him unconscious for days (Xinhua, 2011). Many citizens urged the police to bring the suspect to justice. Police turned to microblogs and released the man's photograph, seeking tips that they hoped would lead to the arrest of the suspect (Xinhua, 2011).

Police in Shenzhen also relied on the Internet in similar protests, resulting in five people turning themselves in. The Fujian district public security bureau posted photos on Sina Weibo of 20 people suspected of destroying public and private property during anti-Japanese protests. Only a day after the photos were posted, authorities had received more than 350 phone calls offering information about the suspects ("China: Five surrender to police for vandalism during anti-Japan protests," 2012).

China has an interest in stirring up anti-Japanese sentiment ahead of the once-in-a-decade leadership transition; occurring at a time of slowing economic growth. Chinese leaders know that nationalism is a useful political tool, and thus uses the Island dispute as a means to do so ("East Asia Territorial Disputes: Geopolitical Implications" 2012). According to a New York Times article,

*Protests in more than 80 Chinese cities, including in urban centers where Japanese car dealerships and electronics plants were damaged, suggested that the Chinese leadership approved the outpouring of nationalism in part as insulation against criticism of the party itself during the transition of power that is scheduled to take place at the 18th Communist Party Congress, set to begin on Nov. 8. But the protests threatened to turn against the Chinese government itself, diplomats and analysts said. (China's Territorial Disputes, 2012)*

Although large-scale protests are rare in China, critics claimed that the government was responsible for organizing and encouraging the protests (Makinen, 2012). The LA Times reported, "last week's anti-Japan demonstrations in China have been a spectacular display of just how easily the ruling Communist Party can harness the power of protest" (Demick and Makinen, 2012). Dissident artist Ai Weiwei said "[i] t is obvious that this was planned," and referenced the Tiananmen Square protests as "the last time that the people themselves organized a real protest and then the government sent in tanks to crush them" (Demick and Makinen, 2012). While there is no evidence that police officers or officials participated in violent acts, there are reports that in a number of cities they directed the public on where to protest and helped clear the streets ("China's Island Disputes," 2012). One police officer posted on a microblog that he needed "to lead the crowd and guide them to march in an orderly fashion" (Demick and Makinen, 2012). The post was later removed. Furthermore, the New York Times reported that approval for the protests had been discussed at very high levels of the government ("China's Territorial Disputes," 2012). Many protestors said that they had been given the day off to demonstrate. Blogger Wen Yunchao said that the government had a hand in organizing the protests, aiding over 500 farmers to travel to Beijing in order to protest (Demick and Makinen, 2012). However, Wen also cautioned that the government was taking a lot of political risks in allowing the protests, because "[n]ot only were there pro-democracy demands, [there were] extreme leftists out demonstrating and those people are even more dangerous to the current government" (Demick and Makinen, 2012).

Some of these protesters were found in Shenzhen and Guangzhou, and were protesting issues beyond the islands, including corruption and high housing prices (Makinen, 2012). One photo circulated on Weibo showed a man wearing a shirt reading, "I'm willing to feed the corrupted officials and become a housing slave, but I will never give up on the Diaoyu Islands" (Makinen, 2012). However, the government has tried to portray the protests as a one-issue cause (the Island dispute).

The government began to fear that the anti-Japanese protests of September 2012 which "emboldened protestors" could turn on the government. Thus, CCTV began to encourage and promote "patriotic" and "rational" protests (Zhai, 2012). Mainland media, including The Beijing News, The China Youth Daily, Shenzhen's

Daily Sunshine, and The Southern Metropolis News all issued statements or ran stories condemning violent protests and calling for a more "mature" and "rational" approach. The Southern Metropolis News warned that if the Chinese "society's passions were aroused too much, social order would collapse" (Zhai, 2012). The LA Times also reported that the protests were beginning to push authorities "toward the edge of their comfort zone," and reported on the calls for rationality (Makinen, 2012). Jessica Chen Weiss, an assistant professor of political science at Yale University, explained that the spread of Weibo and other technology adds a certain amount of unpredictability. She said, "[t]he increasingly viral methods of organizing protests means the government's task [in controlling the demonstrations] is harder" (Makinen, 2012).

This case illustrates that while the Chinese government had the ability to censor social media to inhibit organization of protests by the Chinese, they instead used social media to help serve its interests. In particular, the government turned to social media to aid in the capture and arrest of violent protestors while also allowing these protests to fuel nationalist sentiment against the Japanese at a critical time during a historical transfer of power. Furthermore, evidence suggests that local police forces and other governmental offices aided citizens in their protest efforts. However, as citizens continued to fuel the protests, they began to get out of control, especially in Guangdong province. As such, government officials, who used social media to organize the protests, also used social media as a tool to find those committing violent acts during the protests.

## Wenzhou Train Wreck

On Saturday July 23, 2011, two high-speed trains collided on a viaduct outside of Wenzhou. Train D3115 was traveling from the city of Shaoxing to Wenzhou and came to a halt on the tracks, reportedly because it lost power after being hit by a lightning strike (Ni and Zhang, 2011). Shortly after the loss of power train D3115 was rear ended by train D301 (Ni and Zhang, 2011). The impact derailed two carriages and sent four carriages plummeting off the viaduct, killing 40 people and injuring more than 190 (Dean and Page, 2011). The day after the crash construction machinery was seen demolishing and pushing wrecked rail cars into pits. During a press conference the following Sunday a spokesman for the Railways Ministry explained that by burying the wrecked carriages, room was being made for rescue equipment. However, this answer was met with skepticism, as some believed the true motive was to get the railways operating again (even before all the survivors had been accounted for). Witnesses to the crash site reported that no one seemed to be searching for clues as to what caused the wreck, stating "[n]ot a single person appeared to be photographing or measuring the wreckage or documenting its condition in any way" (Clem, 2011). After rescue efforts ceased, police reportedly continued searching through one badly damaged carriage, finding a two-year-old survivor hours after the railway ministry had said there were no more signs of life.

Premier Wen Jiabao visited four days after the crash, and state journalists questioned him about why all evidence of the accident had been removed from the crash site. Many speculated that the carriages were buried in order to hide evidence of corners being cut or to hide evidence of technology lifted from foreign companies. In fact, 98% of respondents in an unscientific poll on Sina Weibo stated that they believe the reason for burial of the carriage was to destroy evidence (Flanagan, 2011). Initial reports claimed that train D3115 had lost power and stalled after a lightning strike crippled its operational systems. The train's black box recorder was retrieved and the ministry used it to investigate the cause of the crash. In a government report issued in December of 2011, flawed equipment and procedures as well as corrupt practices during the development of China's high speed rail system were the root causes of the Wenzhou train crash. The report found that the lightning strike damaged track-signaling equipment which then failed to send messages between trains allowing for the high-speed train adequate time to stop (Areddy, 2011).

In the immediate aftermath of the wreck, the Department of Propaganda issued a memo to state-run media outlets not to report on the disaster. "Do not question, do not elaborate," it read. "When newspapers came out the next morning, China's first high-speed train wreck was not on the front page" (Osnos, 2012). The dominance of political power over social power has been found to be manifested in the competition to control the narrative of the event and the interaction with society (Liu et al., 2016).

The Chinese government has a vested interest in the railway. The Railway Ministry is manufacturer, operator and regulator of the network, which China began building less than a decade ago, using imported technology from various countries. However, since then, Chinese companies have integrated imported technology into Chinese designs and have been trying to sell Chinese-made high-speed rail equipment overseas. The network is already the world's largest and is planned to stretch some 16,000 kilometers (about 10,000 miles) when it is completed in 2020, at an estimated total cost of more than $300 billion (Dean and Page, 2011). Before the Wenzhou crash, the spokesman for the Railways Ministry bragged that China's technology was so superior to Japan's famous Shinkansen that they "cannot be mentioned in the same breath" (Dean and Page, 2011). However, Japan's high speed railway line has only suffered one casualty in forty-seven years (a passenger who got caught in a doorway) as compared to China, where numerous accidents have claimed dozens of lives ("Dissent in China: Of development and dictators," 2011).

The crash also occurred at an inconvenient time for Party officials, just ahead of the once-a-decade leadership change, in which seven of the nine members of the Politburo Standing Committee will retire. Dean and Page argue that "Party leaders are anxious to avoid the fate of Chinese emperors who were traditionally deemed to have lost the "mandate of heaven"—the divine right to rule—in times of national crisis." While they also maintain that the Wenzhou train crash is not a direct threat to the regime, they also point out that the online media storm highlighted the weaknesses of China's current leadership, "whose rampant corruption and reflexive secrecy could undermine its rule in the long term" (Dean and Page, 2011).

China's railway system appears to be a part of the rampant corruption. In February 2011, Liu Zhijun, the Railways Minister since 2003 and high-speed rail's chief champion, was fired on allegations of "severe violation of discipline," Party slang for "graft". A month later, the Railways Ministry's second-ranking engineer, Zhang Shuguang was dismissed on similar charges.

Party leaders have long viewed management of natural or man-made disasters as vital to Party survival ("Dissent in China: Of development and dictators," 2011). Successful management of a crisis would reflect well on the Party, keeping the public happy. However, the Wenzhou train crash calls into question the Party's ability to manage disasters effectively. Unlike other disasters in China's history (Sichuan earthquake, for instance) party leaders could not distance themselves from the cause of the disaster or shift the blame to local officials. The leaders, advocates of the high-speed rail, had "pinned their own and the country's prestige to high-speed rail," and so the public is holding Beijing responsible for the crash. The link between Beijing and the railway is unavoidable, as just a few weeks before the Wenzhou crash, Beijing officials were touting the opening of a new line between Beijing and Shanghai (which was timed to coincide with the 90th anniversary of the Party founding).

Many citizens expressed concern that the whole disaster may have been caused by the government's reckless pursuit of rapid economic growth with little regard to safety. Five months after the crash, reports released by the government blamed technical failure of an electronic signal that was not adequately tested before being awarded the contract for the construction of the high-speed rail system (Yan, 2012). The report examined the procedures put in place during the bidding process and found corrupt practices used to award construction bids for the project (Areddy, 2011). After a large public outcry, the government ordered the carriages to be dug up and agreed to almost double compensation for the family of each victim to 915,000 yuan ($140,000) (Dean and Page, 2011). Michael Anti, a popular blogger, wrote of the train wreck just after it occurred, "I have never seen this. Yesterday, the train was derailed, today the compartments are buried. Every time there is a plane or train crash, the debris is usually collected and checked to find the cause. Is the Ministry of Railways going nuts?" (Mong and Gu, 2011). One user posted on Weibo, "Can't you just gather the passengers' belongings and wait for their relatives to collect them instead of just burying them? There may be clothes they bought together, pictures they took when they travelled together. In the cell phones, there may be text messages they didn't want to delete, video to record their lives, or even the last words they said before they died!" (Mong and Gu, 2011).

The Wenzhou train disaster brought to light the problem facing Communist Party rule. The Party relies upon high-speed rail growth to create jobs and to keep social tensions in check by providing efficient mass transportation. However, rapid growth has created regime-threatening risks, including deadly accidents such as the Wenzhou train accident and the exposure of corruption at ministerial levels of the government and party.

Dean and Page report that, "Weibo and sites like it have become like a national nervous system, transmitting pulses of opinion and news almost instantly to networks of friends and 'followers'" (2011). Another website also used to transmit news of the train wreck was Tudou, a well-known brand in China that provides a service similar to YouTube (Taylor, 2012). Some users posted images of the train crash online. These social networking sites "played a huge role in exposing the government's cack-handed response" to the Wenzhou train accident ("Asia: A new third rail; A train crash in China," 2011). Many online users called for the dismissal of the current railway minister, Sheng Guangzu. Also, a video circulating online shows what some viewers say were two bodies falling out of carriages as they were being moved away from the rail, prompting further outcry of government mismanagement.

Although the government censors much content on the Internet, it has allowed a surprising degree of openness on Weibo and other sites. Experts argue that the Party sees online activity as a release valve for the public. However, at the same time the government fears what would happen if it took away those outlets.

Dean and Page (2011) argue:

*The new technology has fundamentally changed the relationship between China's government and the governed. The extensive controls on information in the pre-Web days meant that most Chinese were aware of corruption only among local officials, who were convenient scapegoats for national leaders. Now Internet users are aware that issues like food safety problems, land-use abuses and corruption are pervasive.*

Still some commentary was removed by the censors following the crash, including a blog post by Han Han, one of China's most popular bloggers. His essay, "The Derailed Country," criticized the Party for what he called a "heartless approach to development." While the censors often try to block certain key words, the authorities tried to resist doing so during the outbreak of criticism following the Wenzhou train crash.

Many expressed outrage that the Ministry of Railways concluded that the accident was caused by lightning after only two days of investigation. He Weifang, a professor at the Peking University Law School, via his posts on Weibo, brazenly called for a committee to be established to investigate the crash (Ni, Kong, and Zhang, 2011). Some view the use of the internet to criticize the secretive manner in which the government handled the train accident as good, stating that "[t]he one bit of hope for this nation is that, with the advent of the internet, the weak are beginning to find a voice" (Clem, 2011).

According to the Xinhua news agency, following the crash, "the Sichuan Cellphone Press, posted on its microblog a message referencing the power citizens have harnessed by becoming active participants in the creation and dissemination of news and information in recent years" (2011). In fact, the first published picture of the crash site came from a resident of Wenzhou, not the official state media. Microblogs were also used to spread information about the need for blood donors.

Although most of the public fury was online, it also spilled over into mainstream media. Following the accident, Qiu Qiming, a popular anchor on the state broadcaster's CCTV 13 channel appealed to the government, mentioning several safety concerns linked to government mismanagement. On July 29th, despite Premier Wen's calls for "openness and transparency," the Central Propaganda Department issued three sets of prohibitions, essentially forcing the media to cease reporting on the accident, other than stories that accentuated the positive, such as tales of heroism or sacrifice in the relief efforts. Critics argued that Wen's openness and transparency was unlikely to happen, saying that "[Wen] advocates transparency in his public statements, but only insofar as it doesn't threaten the authority of the party" (Wines et. al, 2011). Numerous media outlets were forced to remove or cut down reports about the accident, or run Xinhua Agency reports only:

*The ban issued by the Central Propaganda Department says that "given the increasing complicated nature of public opinion at home and abroad concerning the particularly major Ningbo-Wenzhou railway incident on 23 July, media outlets in various regions, including sub-newspapers, sub-magazines and their news websites, have to immediately bring down the temperature of related reports. Apart from positive reports and news released by authoritative departments, no further report can be released and no further comment made." Major websites have also been asked to remove special reports or move them away from prominent positions on the websites ("Many media outlets defy media gag on Wenzhou train crash reporting," 2011).*

However, some state publications defied orders and ran front page reports or even published special reports. The Global Times, a Beijing newspaper, reported that public opinion could no longer tolerate the arrogance of the railway ministry and others. It reported "[t]he relationship between the government and public is like that of a ship and water: Water can keep the ship afloat or sink it." Journalists and editors unhappy with the gag order turned to online microblogs to voice opinions about the crash.

A year after the Wenzhou train crash, the mainland censors issued a ban on reporting anything related to the train collision, perhaps in the hopes of preventing more public outcry. Many of the victim's families, survivors, or rescuers report that promises made by all levels of the government have not been fulfilled (Yan, 2012).

Response to the train crash created an "online media storm, driven by China's new middle class" (Dean and Page, 2011). A decade ago, it would have been impossible for Chinese to react in such a way because traditional media was tightly controlled by the state, and China had less than 10 million Internet users (Dean and Page, 2011). However, at the time of the Wenzhou train crash, some 485 million Chinese were estimated to use the Internet (Dean and Page, 2011). The most avid online users are young people and members of the urban elite, those that the CCP has had to work harder to woo in recent years. This incident was the first to demonstrate the capacity of both journalists and the public to speak up in such large numbers. In the past, the government could control what journalists wrote and broadcast. Government could and did break up large protests of youth. Recently, however, its control over these political participatory functions has diminished. Social media has given journalists an even larger audience that is less easily controlled by the government and party. Social media has made it easy for youth to disseminate ideas and images. This has led allow citizens to organize multiple areas of protest, making government crackdowns more difficult and risky since these same protestors can use social media to share the government's response with their followers. The harsher the government response, the more it risks escalating tensions around China's most populated cities. Recent events in the Middle East and North Africa dubbed the Arab Spring have now made Chinese leaders acutely aware of the power of social media in organizing and broadcasting government response to such political expression. In fact, there have been similar nascent movements within China called the Jasmine Revolution that have yet to fully take root, but have, nevertheless, required a watchful eye by officials at all levels of government.

The Wenzhou train wreck case demonstrates that social media has the potential to reverse government policy. In the wake of photographs and videos of the train wreckage and the subsequent burial of the carriages, protest leaders took their message to the blogosphere in a plea for change in government actions. Not less than one day after the carriages were buried, they were dug up. Social media is eroding the government's ability to literally bury its corruption and its secrecy. To the extent possible, citizens are attempting to make transparent what the government, or its agencies, are unwilling to do. Additionally, the Wenzhou train wreck case shows, as in the Island Dispute case discussed above, the difficulty the government has in controlling the media as a result of the accessibility and popularity of social media. When the government issues a gag order to silence journalists in state-run media outlets, journalists can now turn to social media to continue to voice their views on the matter and circumvent the state controls to a large extent.

## Wukan Protest

Starting in September 2011, Wukan villagers protested the sale of local agricultural land to developers without their consent. The villagers say that the local government and Lufeng authorities illegally sold the villagers' collectively owned land to developers without their permission. Furthermore, money from the sales was unaccounted for (Wong, 2012). During initial protests, the local party headquarters were raided and six police cars were overturned and windows were smashed. The protests began to spread to neighboring communities (Lua, 2012). A few weeks later, authorities agreed to let Wukan villagers elect thirteen representatives to engage in negotiations. Dissatisfied with developments, in November, over 5,000 villagers marched more than seven kilometers to the county-level government office to protest, but dispersed after government officials promised to address their concerns.

The conflict peaked in December when key village representatives were taken into police custody. Two days later, 42-year-old villager, Xue Jinbo, one of the twelve people chosen by villagers to negotiate a settlement to the land dispute, died. Xue was elected the head of a provisional village committee after officials fled Wukan (The Nation, 2012). Lufeng police maintain that the men were arrested and charged with protest-related crimes and that Xue died of a heart attack. However, relatives of Xue who were allowed to see Xue's body said that there was caked blood and bruises on him and he had a broken left thumb (Wong, 2011). His death incited more riots within the village. Villagers forced police and government officials to flee Wukan. The police surrounded the village, cutting off supplies. The villagers responded by barricading the roads into the village with tree trunks and set up armed guard posts. Villagers managed to get supplies through using only narrow paths. Following this uprising, the government began to censor the Internet blocking any information referring to the developments in Wukan (Lua, 2012).

However, despite tight media control, news of the protests broke out within days of government officials fleeing Wukan. Photographs posted on Chinese websites showed thousands of villagers holding banners and marching at the village square (Krishnan, 2011). A photograph of Xue was plastered on the empty police station's front gates. Revolutions in the Arab world have raised concerns within and outside of China worrying government officials if similar protests could occur in China (Yongnian, 2012). The New York Times reports that stories about the Wukan protest were banned from all Chinese media and Internet sites. However, there were signs that word of the protests was spreading. Footage posted online shows Wukan village protesters ransacking police vehicles, while police kick and beat villagers before apparently being driven away (The Nation, 2012). Users on Weibo reported land dispute protests in three other villages in Shanwei Prefecture. In Guangzhou, three people were arrested after a protest in sympathy with the Wukan villagers. Another microblog post included photographs and a description of a violent clash between police officers and thousands of people in Haimen, a township in Shantou. The microblog posts said that the Shantou demonstrators were protesting plans to build a power plant, fearing that it would add to pollution and damage the local fishing industry (Wong, 2012).

Fear of news about the Wukan village uprising prompted the censors to block the terms "Wukan" and "Xue Jinbo" (The Nation, 2011). However, citizens continued to post images of villagers with bloodied faces and bruised bodies (The Nation, 2012). Citizens have developed different words or used characters in order to get around the censors. For example, Wukan is commonly referred to as the "W village" or Niaoqian village (written in characters) in order to avoid the censors (Lua, 2012). In a webpost once deleted by Chinese government censors and reposted as an image in a microblog, Chinese journalist Chen Jibing discussed corruption as the root cause of the Wukan protest, while pointing to specific people and acts of corruption. Chen writes, "While for reasons known to all newspapers, television and other mass media kept quiet on this incident out of fear, it was the most hotly watched public opinion storm on China's internet — and particularly on microblogs — for some time" (Bandurski, 2012).

In response to citizens finding ways to subvert government directives using social media, and growing unease among central government leaders, Beijing has prompted authorities to issue a controversial new rule that would require Weibo users to register their real names before they could post online (The Nation, 2012). In this way, Beijing hoped to remove the veil of anonymity that citizens enjoyed while trying to lift the government's veil of secrecy.

Eventually, village representatives and provincial officials came to a deal with the aid of Wang Yang, the Guangdong party secretary. Rather than end the standoff with violence, Yang conceded that the villagers had legitimate concerns and promised an investigation and elections to replace the local officials who fled Wukan during the thirteen-day standoff. The March 2012 elections filled a seven-member committee, which included a village chief and two deputies (Lau, 2012). The elected officials exercised power over local finances and the sale and apportioning of farmland.

Villagers described the election "as the first real poll in years" (Trayer, 2012). However, Guangdong party secretary Wang Yang said that the Wukan election was not political reform, but instead a return to the rule of the Organic Law, which allows the election of local officials but, according to Yang, had not been practiced in previous Wukan elections. By casting past elections as corrupt, Yang was able to argue that the Wukan elections did not represent "an innovative model for democracy," but instead a return to party approved practices (Trayer, 2012).

The Wukan case highlights the double-edged sword that social media presents to Chinese leadership. Central government officials have an interest in weeding out corrupt local government officials. However, if corrupt practices go unaddressed by the central or provincial level governments, citizens-turned-protest-leaders pound the keyboard (or cell phones) to share information and images of the developments on the ground in an attempt to organize like-minded citizens to support their cause. Social media in this respect challenges the ability of the government, both at the central and provincial levels, to control the situation to find a solution before local government officials and police forces get run out of the village. Once again, as in the preceding two cases, we see that government attempts to control the media in China are undermined by journalists posting information on websites and social media microblogs, linking developments on the ground to issues in cyberspace. As government censors put tougher restrictions on what people can post, citizens continue to find innovative and creative ways to subvert these censors to get their message posted, thereby creating links between cyberspace and real space.

## THE FINDINGS AND RECOMMENDATIONS

These case studies illustrate the dilemma that Chinese government officials face with respect to social media. Social media helps the government pursue some of its interests at times, such as increasing nationalism and support for the party during the island dispute case. Social media also helps them to gauge and connect with public opinion, which in turn allows central leaders to put pressure on local leaders, as in the case of Wenzhou train crash and Wukan uprising. Additionally, as seen in the Sichua/Wenchuan Earthquake, Wenzhou and Wukan cases, social media and the resulting pressures caused by public outcry can lead to the exposure of corruption even in high levels of the government. However, as is illustrated in the Wukan case, central government leaders kept a close eye to see whether or not the outcome of the village of Wukan would lead to similar uprisings around the country. The government closely monitors any social mobility that was seen as part of a Jasmine Revolution in China during the winter of 2011. Lastly, as is illustrated in the Sichuan/Wenchuan, Wenzhou and Wukan cases, and detailed in the King et al. study, the Chinese government is censoring speech less and less on the various social media outlets and increasingly trying to concentrate their efforts on restricting speech related to the mobilization of people. That is to say, more and more speech critical of the government and its policies is able to be disseminated via social media, but the government continues to block, to the extent possible, speech related to the organization of masses.

*Contrary to previous understandings, posts with negative, even vitriolic, criticism of the state, its leaders, and its policies are not more likely to be censored. Instead, we show that the censorship program is aimed at curtailing collective action by silencing comments that represent, reinforce, or spur social mobilization, regardless of content. Censorship is oriented toward attempting to forestall collective activities that are occurring now or may occur in the future --- and, as such, seem to clearly expose government intent. (King et al., in press).*

The findings above imply a series of growing tensions between the state and society. The Chinese state is trying to use social media to pursue its ends and to avoid a popular backlash as it tightens controls over social media. There is tension between the states' attempts to control social media and the growing demands by citizens to more and more unfettered use of this technology as well as in citizen's increasingly creative methods at subverting these government controls which in turn increases their ability to use social media for their own ends. Tension exists between the state's desire to control the policy making process and society's desire for a greater role in the policy process in China through the dissemination of information aided by their increasing ability to mass organize. These tensions are not likely to ease any time soon as Chinese officials continue to crimp access to information when trouble breaks out, while more and more Chinese citizens are accessing, creating, and innovating new information, making it more difficult for the government to be successful in blocking this information.

As pervasive corruption affects more Chinese citizens their demands for change will likely grow. However, the government's response will, as has generally been the case in the past, continue to be evasive and veiled in secrecy. To the extent that the government is able to persuade people that their economic well-being depends largely in part on social stability, local problems and controversies exposed by citizens, even with widespread coverage by social media, are not likely to bring about significant national change within China anytime soon. This means that changes within the current one-party rule structure are not likely to occur, while expansions of political and social rights for the citizens will remain frozen for the foreseeable future.

## FUTURE RESEARCH DIRECTIONS

Access to reliable information from China has been difficult because lower level government officials have great incentives to report positive information up stream to the central government, with most of the manipulation of data taking place at the municipal and provincial levels. The incentives come mainly in the form of promotions within the party based on economic success at these lower levels. Mass organization and protest, among other disturbances can be aided by communication via social media. For government officials, these disturbances can lead to greater levels of instability which in turn can negatively affect economic development.

Interviewing policy makers about the influence that public demonstrations and public outcry on various social media outlets has on policy would offer insight into how these public displays affect the policy formation process. However, interviewing elites at the top echelons of power within the Chinese Communist Party about the influence of social media on policy within the state is virtually impossible. As such, this research relied primarily on secondary sources. Future research should aim to overcome obstacles to elite interviewing to explore a single event and the influence that social media may or may not have had on policy outcomes. With interviews of elite officials of central, provincial and municipal governments we could better understand the nature of the double-edged sword: when officials are likely to allow citizens to vent frustrations via social media in support of state interests and when officials are likely to block the transmission of information via social media because of a perceived threat to the state. Interviewing individuals involved in the events and microblogging posts would help to connect social media more directly with policy changes over time. Repeating this model in other instances would allow us to better understand the conditions under which social media is likely to have a more significant impact and the conditions under which it is likely to have little influence. In this way, we will learn just how sharp each side of the sword is.

## REFERENCES

Abbott, J., MacDonald, A., & Givens, J. (2013). New Social Media and (Electronic) Democratization in East and Southeast Asia: Malaysia and China Compared. *Taiwan Journal of Democracy*, *9*(2), 105–137.

Activists in China's Wukan village being "spied upon, harassed". (2012, March 10). *BBC Monitoring Asia Pacific-Political.*

Ai, W. (2018) Ai Weiwei: The artwork that made me the most dangerous person in China. *The Guardian.* Retrieved from https://www.theguardian.com/artanddesign/2018/feb/15/ai-weiwei-remembering-sichuan-earthquake

Alexander, A. (2011, February 9). Internet Role in Egypt's Protests. *BBC News.* Retrieved from: http://www.bbc.co.uk/news/world-middle-east-12400319

Amnesty International. (2009) Justice Denied: Harassment of Sichuan earthquake survivors and activists. *Amnesty International Publications.* Retrieved from https://www.amnesty.org/en/documents/ASA17/018/2009/en/

Anti-Japan protests continue in China over territorial row. *BBC Monitoring Asia Pacific-Political.*

Areddy, J. (2011, December 29). China Blasts High-Speed Rail System. *The Wall Street Journal.* Retrieved from: http://online.wsj.com/article/SB10001424052970204632204577126121683353312.html

Arthur, C. (2011, January 28). Egypt Cuts Off Internet Access. *The Guardian.* Retrieved from: http://www.guardian.co.uk/technology/2011/jan/28/egypt-cuts-off-internet-access

Asia: A new third rail. (2011, July 30). *The Economist.* Retrieved from: www.economist.com/node/21524914

Bandurski, D. (2012, January 30). The Lagacy of Wukan. *China Media Project.* Retrieved from: http://cmp.hku.hk/2012/01/30/18237/

Bandursky, D. (2009). Looking Back on Chinese Media Reporting of School Collapses. *China Media Project.* Retrieved from http://chinamediaproject.org/2009/05/07/one-year-of-chinese-media-reporting-on-school-collapses/

Blight, G., Rininsland, A., Rogers, S., & Torpey, P. (2012, April 16). How Open is Your Internet? An interactive map. *The Guardian.* Retrieved at: http://www.guardian.co.uk/technology/datablog/interactive/2012/apr/16/web-filtering-censorship

Bradsher, K., & Fackler, M. (in press). Protesters across China vent anger at Japan; Dispute over island chain adds fuel to tensions as nationalism is inflamed. *The International Herald Tribune.*

Breslin, S. (2008). *Democratising One-Party Rule? Political reform, national, and legitimacy in the People's Republic of China* (Working Paper no. 67). Retrieved from FRIDE website: www.fride.org/download/wp67_china_one_party_eng2_sep08.pdf

Bristow, M. (2009, November 23). Prison term for Chinese activist. *BBC News*. Retrieved from http://news.bbc.co.uk/go/pr/fr/-/2/hi/asia-pacific/8373573.stm

Buckley, C. (2008, June 16). China prosecurtors join quake school collapse probe. *Reuters*. Retrieved from https://www.reuters.com/article/idINIndia-34078920080616

Chen, Y. (2013). *Public Opinion Guidance of Wechat Dissemination in Major Events*. Retrieved from http://www.cqvip.com/qk/81149x/201413/661594554.html

China: 5,335 students killed or missing after 2008 quake. (2009, May 10). *CNN*. Retrieved from http://www.cnn.com/2009/WORLD/asiapcf/05/07/china.quake.deaths/index.html

China feels the quake online. (2008, May 16). *BBC News*. Retrieved from http://news.bbc.co.uk/2/hi/asia-pacific/7402612.stm

China: Five surrender to police for vandalism during anti-Japan protests. *BBC Monitoring Asia Pacific-Political.*

China Internet Network Information Center. (2012, January). *Statistical Report on the Internet Development in China* (Abridged Edition). Retrieved at: http://www1.cnnic.cn/IDR/ReportDownloads/201209/P020120904421720687608.pdf

China-Japan dispute. (2010, September 26). *The Korea Herald.*

China sacks railway spokesman over remarks on train crash. (n.d.). *BBC Monitoring Asia Pacific-Political.*

China Struggles to Tame Microblogging Masses. (2011, September 8). *The Independent.*

China's Territorial Disputes. (2012, October 12). *The New York Times*. Retrieved from http://topics.nytimes.com/top/news/international/countriesandterritories/china/territorial-disputes/index.html

Chinese man held over earthquake photos. (2008, July 31). *The Telegraph*. Retrieved at https://www.telegraph.co.uk/news/worldnews/asia/china/2477024/Chinese-man-held-over-earthquake-photos.html

Chinese web users condemn video of protester beating owner of Japan-made car. *BBC Monitoring Asia Pacific-Political.*

Chung, J. (2011). *Weibo and 'Iron Curtain 2.0' in China: Who Is Winning the Cat-and-Mouse Game?* East Asia Institute Issue Briefing, No. MASI 2011-07. Retrieved at: http://www.eai.or.kr/type/panelView.asp?code=eng_report&catcode=&idx=10700&bytag=p

Clem, W. (in press). Seeing through all the talk of transparency; An 'open, transparent' investigation in the Wenzhou train crash? We've heard that kind of talk before. *South China Morning Post.*

Coonan, C. (2009, May 4). China's quake cover-up. *The Independent.* Retrieved from https://www.independent.co.uk/news/world/asia/chinas-quake-cover-up-1678542.html

Coonan, C. (in press). Wukan villagers end protest after getting key concessions. *The Irish Times.*

Dean, J., & Page, J. (2011, July 30). Trouble on the China Express. *The Wall Street Journal.*

Demick, B. (in press). China-Japan island dispute fuels protests; Nationalist fervor by activists leads to vandalism and efforts to reach one isle. *Los Angeles Times.*

Demick, B., & Makinen, J. (in press). Beijing's hand seen in anti-Japanese protests; Officials appeared to have guided marchers, but then some rallies went out of control. *Los Angeles Times.*

Deng, Y., & Moore, T. G. (2004). China Views Globalization: Towards a New Power Politics. *The Washington Quarterly*, *27*(3), 115–136. doi:10.1162/016366004323090296

Diaz, L., Hart Dubner, B., & Parent, J. (2007). When is a "rock" an "island"? – Another unilateral declaration defies "norms" of international law. *Michigan State Journal of International Law,* 519-555.

Dissent in China: Of Development and Dictators. (2011, August 6). *The Economist.* Retrieved at: http://www.economist.com/node/21525419

East Asia Territorial Disputes. Geopolitical Implications. *Business Monitor Online.* Retrieved October 29, 2012, from http://wwws.moreheadstate.edu:2052/hottopics/lnacademic/?

Eimer, D. (in press). A new Orwellian euphemism for stifling online dissent; David Eimer says 'cultural security' targets the messengers, not the issues. *South China Morning Post.*

Five Years after a Quake, Chinese Cite Shoddy Reconstruction. (2013, May 13). *National Public Radio* retrieved from https://www.npr.org/sections/parallels/2013/05/14/183635289/Five-Years-After-A-Quake-Chinese-Cite-Shoddy-Reconstruction

Flanagan, E. (2011, July 29). China's All Powerful Rail Ministry Takes Heat for Crash. *NBC News.* Retrieved from http://behindthewall.nbcnews.com/_nv/more/section/archive?date=2011/7

Ford, P. (2010, February 9). China sentences quake activist Tan Zuoren. *Christian Science Monitor.* Retrieved from https://www.csmonitor.com/World/Asia-Pacific/2010/0209/China-sentences-quake-activist-Tan-Zuoren

Gainous, J., Wagner, K., & Abbott, J. (2015). Civic Disobedience: Does Internet Use Stimulate Political Unrest in East-Asia? *Journal of Information Technology & Politics, 12*(2), 219–236. doi:10.1080/19331681.2015.1034909

Griggs, B. (2013, March 10). Beyond the DMZ: Tweeting from North Korea. *CNN.* Retrieved at: http://www.cnn.com/2013/03/10/tech/social-media/social-media-north-korea/index.html

Gross, D. (2012, December 3). Syria Caused Internet Blackout Security Firm Says. *CNN.* Retrieved at: http://www.cnn.com/2012/11/30/tech/web/syria-internet/index.html

Hughes, C. (2000). Nationalsim in Chinese Cyberspace. *Cambridge Review of International Affairs, 13*(2), 195–209. doi:10.1080/09557570008400309

In China anti-Japan protests over disputed islands get some zing. (2012, Sept. 13). *L.A. Times.*

Information Office of the State Council of China. (2008, July 6). *Wenchuan Earthquake has Killed 69,196 People and Left 18,379 Missing.* Retrieved from http://news.sina.com.cn/c/2008-07-06/162615881691.shtml

Jiao, P. (in press). Party chief rant lights debate; One city official has brought on praise and mockery over remarks to villagers protesting against land seizures. *South China Morning Post.*

Jiao, P. (in press). Weibo puts the heat on officials over tragedy; Despite propaganda to downplay a high-speed train crash, microblog site and media force authorities to give more accurate information on the accident. *South China Morning Post.*

Jing, M., Zhou, Y., & Ma, D. (2014). *The Communication Mode, Characteristics and Reflection of WeChat.* Retrieved from https://wenku.baidu.com/view/054a486fe55c3b3567ec102de2bd960590c6d91a.html

Kaufman, S. (2008, May 20). *China's Government Praised for Easing Media Restrictions*. Retrieved from https://web.archive.org/web/20080703063114/http://www.america.gov/st/freepress-english/2008/May/20080520153034esnamfuak0.7039911.html

Khamis, S., & Vaughn, K. (2011). Cyberactivism in the Egyptian Revolution: how civic engagement and citizen journalism tilted the balance. *Arab Media & Society*, 14. Retrieved at: http://www.arabmediasociety.com/?article=769

King, G., Pan, J., & Roberts, M. (In press). How Censorship in China Allows Government Criticism but Silences Collective Expression. *American Political Science Review*. Retrieved at: http://gking.harvard.edu/files/gking/files/censored.pdf

Krishnan, A. (2011). A Unique Protest in China. *The Hindu*. Retrieved at: http://www.thehindu.com/news/international/a-unique-protest-in-china/article2715292.ece

Kuhn, A. (2008, May 4). Chinese seek answers in student quake deaths. *National Public Radio*. Retrieved at https://www.npr.org/templates/story/story.php?storyId=103727282

Kwong, P. (2009, May 11). Silencing the Messenger. *New York Times*. Retrieved from https://www.nytimes.com/2009/05/12/opinion/12iht-edkwong.html?searchResultPosition=10

LaFraniere, S. (in press). A risky about-face in China; Yielding to the merit of the Wukan protests, politician defuses revolt. *The International Herald Tribune*.

LaFraniere, S. (in press). Media Blackout in China After Wreck. *The New York Times*.

Lai, A. (2012, April 4). Chinese artist Ai Weiwei places himself under home surveillance. *CNN*. Retrieved from https://www.cnn.com/2012/04/04/world/asia/ai-weiwei-webcams/index.html

Lao, M. (2018, May 10). A Decade-long search for justice on student deaths. *Inkstone*. Retrieved at https://www.inkstonenews.com/china/parents-call-investigation-shoddy-school-buildings-sichuan-earthquake-anniversary/article/2145722

Lau, M. (2011, December 22). Olive branch brings life back to normal in Wukan; Villages remove banners and cancel protest after party officials offer to free detainees and back a committee in dispute over land grabs and death in custody. *South China Morning Post*.

Lau, M. (2012, January 10). Victory over corruption; A protest by villagers over illegal land sales forced the normally hardline mainland authorities to negotiate – and could set a precedent in the future. *South China Morning Post*.

Lau, M. (2012, October 6) Pressure on protest leader turned party chief. *South China Morning Post.*

Lau, M. (2012, October 6). Wukan finds democracy hard going; Guangdong villagers freely elected their leaders after months of protest, but running their own affairs and getting their land back is proving tough. *South China Morning Post.*

Levin, D. (2013, May 11). *Social Media in China Fuel Citizen Response to Quake.* Retrieved from https://www.nytimes.com/2013/05/12/world/asia/quake-response.html?login=email&auth=

Li, L., & O'Brein, K. (2010). Protest Leadership in Rural China. In P. Hays & S. Rosen (Eds.), *Chinese Politics* (pp. 85–108). London, UK: Routledge.

Liu, S., & Chang, T. (2016). One Disaster, Three Institutional Responses: Legitimation crisis and competing discourses in China. *Journalism Studies*, *9*(1), 1–23.

Magistad, M. (2012). How Weibo is Changing China. *Yale Global Online Magazine.* Retrieved at: http://yaleglobal.yale.edu/content/how-weibo-changing-china

Makinen, J. (in press) Isle dispute fuels offbeat protests by Chinese; Japanese vehicles and online porn are being dragged into Beijing's tug of war with Tokyo. *Los Angeles Times.*

Makinen, J. (in press). Anti-Japan rallies break out in more China cities; Violence continues as protesters target diplomatic posts and business in a dispute over uninhabited isles. *Los Angeles Times.*

Manjiao, C. (2011). The Unhelpfulness of Treat Law in Solving the Sino-Japan Sovereign Dispute over the Diaoyu Islands. *East Asia Law Review,* 163-189.

Many media outlets defy China gag on Wenzhou train crash reporting. (n.d.). *BBC Monitoring Asia Pacific-Political.*

Mong, A., & Gu, B. (2011, July 25). Deadly Crash in China Provokes Outrage. *NBC News.* Retrieved from http://behindthewall.nbcnews.com/_nv/more/section/archive?date=2011/7

Moore, M. (2008). China earthquake brings out citizen journalists. *The Telegraph.* Retrieved from https://www.telegraph.co.uk/news/worldnews/asia/china/1950212/China-earthquake-brings-out-citizen-journalists.html

Ni, Y., Kong, X., & Zhang, Y. (2011, July 27). Xinhua Insight: Relatives' sorrow amid claims and doubts regarding China Train Crash. *Xinhua General News Service*. Retrieved at: http://news.xinhuanet.com/english2010/indepth/2011-07/27/c_131013761.htm

O'Brien, K. (Ed.). (2008). *Popular Protest in China*. Cambridge, MA: Harvard University Press. doi:10.4159/9780674041585

O'Brien, K., & Stern, R. (2008). Introduction: Studying Contention in China. In K. O'Brien (Ed.), *Popular Protest in China* (pp. 11–25). Cambridge, MA: Harvard University Press. doi:10.4159/9780674041585-intro

Olesen, A. (2012, August 2). Sina Weibo, China's Twitter, leads microblog craze. *The Huffington Post*. Retrieved at: http://www.huffingtonpost.com/2012/08/02/sina-weibo-chinas-twitter_n_1732300.html?vi

Osnos, E. (2012, October 22). Boss Rail: The disaster that exposed the underside of the boom. *The New Yorker*. Retrieved at: http://www.newyorker.com/reporting/2012/10/22/121022fa_fact_osnos?currentPage=1

Parsons, T., Ji, C., & Kirby, E. (2008). Stress changes from the 2008 Wenchuan earthquake and increased hazard in the Sichuan basin. *Nature*, *454*(7203), 509–510. doi:10.1038/nature07177 PMID:18600260

Perlez, J. (in press). China Accuses Japan of Stealing After Purchase of Group of Disputed Islands. *The New York Times.*

Police break up protest by parents of China earthquake victims. (2008, June 3). Retrieved from https://www.theguardian.com/world/2008/jun/03/chinaearthquake.china?gusrc=rss&feed=networkfront

Protests in Taiwan over Japan's purchase of disputed islands 23 September. (n.d.). *BBC Monitoring Asia Pacific-Political.*

Qiang, X. (2011). The Battle for the Chinese Internet. *Journal of Democracy*, *22*(2), 47–61. doi:10.1353/jod.2011.0020

Qu, Y., Huang, C., Zhang, P., & Zhang, J. (2011). Harnessing social media in response to major disasters. *CSCW 2011 Workshop: Designing Social and Collaborative Systems for China*.

Qu, Y., Wu, P., & Wang, X. (2009). *Online Community Response to Major Disaster: A Study of Tianya Forum in the 2008 Sichuan Earthquake*. Presented at the *42nd Hawaii International Conferences on System Sciences*, Honolulu, HI.

Rapoza, K. (2011). Can China's Renren survive Sina's Weibo? *Forbes*. Retrieved at: http://www.forbes.com/sites/kenrapoza/2011/08/16/can-chinas-renren-survive-sinas-weibo/

Schudson, M. (1995). *The Power of News*. Cambridge, MA: Harvard University Press.

Sichuan earthquake: government prohibits investigations into collapsed schools. (2009, April 24). *Asia News IT.* Retrieve from http://www.asianews.it/news-en/Sichuan-earthquake:-government-prohibits-investigations-into-collapsed-schools-15072.html

Sichuan earthquake killed more than 5,000 pupils, says China. (2009, May 7). *The Guardian*. Retrieved from https://www.theguardian.com/world/2009/may/07/china-quake-pupils-death-toll

Statistical Survey Report on the Internet Development in China, Abridged Edition. (2008, July). *China Internet Network Information Center*. Retrieved at: http://www.apira.org/data/upload/pdf/Asia-Pacific/CNNIC/22rdreport-en.pdf

Taiwan fishermen plan protest trip to disputed islands with Japan. (n.d.). *BBC Monitoring Asia Pacific-Political.*

Taiwan legislature demands firm stance on disputed islands. (n.d.). *BBC News Monitoring Asia Pacific-Political.*

Taylor, B. (2012). *Troubled Waters in the East China Sea*. The Australian.

Teacher, S., & Liu, S. was Released to Serve his Reeducation-Through-Labor Sentence Outside of Labor Camp. (2008, September 26). *Human Rights in China*. Retrieved at https://www.hrichina.org/en/content/220

The Nation. (2011). Mainland protests. *The Nation.*

The Nation. (2012). The China Post. *The Nation.*

Tilly, C., & Tarrow, S. (2007). *Contentious Politics*. Boulder, CO: Paradigm.

Trayer, T. (2012). Wukan holds local elections, displays civil disobedience. *Human Rights Brief,* 32-34.

Trindle, J. (2008, May 29). *China Questions Construction Quality*. Retrieved from https://web.archive.org/web/20080613062138/http://voanews.com/english/2008-05-29-voa12.cfm

Tudou launches punchy Nasdaq IPO, tests investor bid for China. (n.d.). *EuroWeek*.

Tze-wei, N. (in press). Payout offer raised after public outcry; Cash compensation to families who lost relatives in the rail disaster is increased from 500,000 yuan to 915,000 yuan, but some critics say this is still too low. *South China Morning Post.*

Tze-wei, N. (in press). Microblogs' contribution to rule of law praised; Editor of official review of legal developments cites online campaign against child abduction. *South China Morning Post.*

Tze-wei, N. (in press). Survival of two and a half year-old girl moves many hearts. *South China Morning Post.*

Ukkusuri, S., Zhan, X., Sadri, A., & Ye, Q. (2014). Use of Social Media Data to Explore Crisis Informatics: Study of 2013 Oklahoma Tornado. *Transportation Research Record: Journal of the Transportation Research Board, 2459*(1), 110–118. doi:10.3141/2459-13

Video: Tan Zuoren's Sichuan Earthquake Investigation. (2009, December 13). Retrieved from https://chinadigitaltimes.net/2009/12/video-tan-zuorens-sichuan-earthquake-investigation/

Weaver, M. (2008, June 3). *Police break up protest by parents of China earthquake victims.* Retrieved from https://www.theguardian.com/world/2008/jun/03/chinaearthquake.china?gusrc=rss&feed=networkfront

Wen, P. (in press). Purge lifts lid on inner workings of politburo. *Sydney Morning Herald.*

Wen, P. (in press). Wang Yang wouldn't admit it, of course, but he may be the future face of China. *Sydney Morning Herald.*

Weston, T. B., & Jensen, L. M. (2012). *China In and Beyond the Headlines.* Lanham, MA: Rowman and Littlefield.

Winerman, L. (2009). Crisis Communication. *Nature, 457*(7228), 376–378. doi:10.1038/457376a PMID:19158766

Wines, M., Ansfield, J., & LaFraniere, S. (2011, August 8). China's Premier Seeks Reforms and Relevance. *The New York Times.*

Wong, E. (2009, August 12). Chinese Artist Says He Was Barred From Rights Advocate's Trial. *New York Times.* Retrieved from https://www.nytimes.com/2009/08/13/world/asia/13china.html

Wong, E. (in press). Demonstrators Who Took Over Chinese Village Halt Protest. *The New York Times.*

Wong, G. (2010, February 9). Chinese activist who probed collapsed schools in Sichuan quake jailed. *The Toronto Star*. Retrieved from https://www.thestar.com/news/world/2010/02/09/chinese_activist_who_probed_collapse_of_schools_in_sichuan_quake_jailed.html

Wu, J. (2008, May 12). School building quality to be probed. *China Daily*. Retrieved from http://www.chinadaily.com.cn/china/2008-05/17/content_6692329.htm

Wu, J. (2008, May 17). *School building quality to be probed*. Retrieved from http://www.chinadaily.com.cn/china/2008-05/17/content_6692329.htm

Xinhua notes active response of "we-media" to Chinese train crash. (2011, July 25). *BBC Monitoring Asia Pacific-Political.*

Xu, B., & Albert, E. (2017). *Media Censorship in China*. Council on Foreign Relations.

Xu, L. (2008, June 7). *The Power of Chinese Netizens After the Earthquake*. Retrieved from http://english.ohmynews.com/articleview/article_view.asp?menu=c10400&no=382780&rel_no=1&isPrint=print

Yan, A. (in press). Rail-crash relatives are resigned to the outcome; One year after the Wenzhou high-speed train disaster, families continue to mourn but the authorities have failed to answer their grievances. *South China Morning Post.*

Yang, G. (2008). Contention in Cyberspace. In K. O'Brien (Ed.), *Popular Protest in China* (pp. 126–143). Cambridge, MA: Harvard University Press. doi:10.4159/9780674041585-008

Yardley, J. (2008, May 25). *Chinese Are Left to Ask Why Schools Crumbled*. Retrieved from https://www.nytimes.com/2008/05/25/world/asia/25schools.html?action=click&module=RelatedCoverage&pgtype=Article®ion=Footer

Yongnian, Z. (in press). Guangdong's reform and China's future. *The Straits Times.*

York, G. (2016, December 01). *Beijing can't muzzle outrage over deadly collapsed schools*. Retrieved from https://www.theglobeandmail.com/news/world/beijing-cant-muzzle-outrage-over-deadly-collapsed-schools/article674653/

Yu, X. (2015). *Matthew effect, barrel effect and trickle-down effect of Wechat transmission*. Retrieved from http://www.doc88.com/p-2572805204612.html

Zhai, K. (2012, September 23). After the flames died down, some national soul-searching: Protest coverage ran from calls for rational behaviour to tough questions on patriotism. *South China Morning Post.*

Zhang, X. (2011). Mass media in China. In *Understanding Chinese Society* (pp. 156–169). New York, NY: Routledge.

## ADDITIONAL READING

Chung, J. (2008). China and South Korea: The Internet and the Political Regime. *Asian Survey*, *48*(5), 727–751. doi:10.1525/AS.2008.48.5.727

Deng, Y., & Moore, T. G. (2004). China Views Globalization: Towards a New Power Politics. *The Washington Quarterly*, *27*(3), 115–136. doi:10.1162/016366004323090296

Hassid, J. (2008). Controlling the Chinese Media: An Uncertain Business. *Asian Survey*, *48*(3), 414–430. doi:10.1525/as.2008.48.3.414

Hays, P., & Rosen, S. (Eds.), *Chinese Politics*. London, UK: Routledge.

Ho, P. (2012). Embedded Activism and Political Change in a Semiauthoritarian Context. *China Information*, *21*(2), 187–209. doi:10.1177/0920203X07079643

Lee, C.-C. (Ed.). (1994). *China's Media, Media's China*. Boulder, CO: Westview Press.

Lieberthal, K. (2009). *Governing China*. Lanham, MA: Rowan & Littlefield.

O'Brien, K. (Ed.). (2008). *Popular Protest in China*. Cambridge, MA: Harvard University Press. doi:10.4159/9780674041585

Shirk, S. (Ed.). (2011). *Changing Media, Changing China*. New York, NY: Oxford University Press.

Weston, T. B., & Jensen, L. M. (2012). *China In and Beyond the Headlines*. Lanham, MA: Rowman and Littlefield.

Wu, X. (2007). *Chinese Cyber Nationalism*. Lanham, MA: Rowman and Littlefield.

Zheng, Y., & Zhang, X. (Eds.). (2009). China's Information and Communications Technology Revolution: Social Changes and State Responses. London, UK: Routledge.

## KEY TERMS AND DEFINITIONS

**Arab Spring:** Social movement occurring in several North African countries (Tunisia, Libya, Egypt, and Syria, among others) organized in part through the use of the Internet and social media.

**Blogosphere:** The collection of web-based personal logs or journals.

**Cyberactivism:** Political and social activism using the Internet as an aid to organize and disseminate information.

**Jasmine Revolution:** Nascent social movement in China in response to the Arab Spring.

**Legitimacy:** Popular belief that the current government and its leaders are the "rightful" ones to govern.

**Microblog:** User created weblog that is limited to a couple of hundred characters.

**Protest Leader:** Those who initiate petitions, organize mass demonstrations and other types of collective action that target political leaders and policies.

**Social Media:** Media based on the Internet allowing users to write stories, upload pictures and video, and contribute to others' postings.

# Chapter 8
# Electoral Reform and the Electronic Voting System:
## Case of Mongolia

**Dashbalbar Gangabaatar**
*Nagoya University, Mongolia*

## ABSTRACT

*Mongolia introduced a new electronic voting system for the first time for the 2012 parliamentary election. E-voting empowers citizens by making voting simpler and providing better opportunities for certain groups of citizens to participate in the election process. The electoral reform was one of the major steps the parliament carried out in order to restore public trust lost in the violent protests against the 2008 parliamentary election results. A free, transparent, and fair electoral system was important to correct the fraud in the old election system. This chapter examines the effectiveness of the mixed system of election, the electronic voting system, the constitutionality of the electoral systems, and other changes to the electoral system in Mongolia.*

## INTRODUCTION

The Election law of Mongolia was revised in December 2011 to introduce new changes to the electoral system. Instead of the majoritarian system a parallel system, which uses elements of both majority and proportional systems running independently of each other, was adopted. Under the revised law, 28 of the 76 members of the State Great Khural will be elected under a proportional representation system while the remaining 28 members will be chosen by a majority vote within single-member

DOI: 10.4018/978-1-7998-3152-5.ch008

districts. Among other changes, the introduction of electronic voting and the use of other modern technologies in parliamentary election were crucial for the conducting of a fair, transparent and legitimate election.

In previous elections, there have been numerous allegations of election frauds, of which only a small number of allegations were documented and concrete evidence was provided. However, after the 2008 parliamentary election, protesters dissatisfied with the election results and the alleged fraud gathered at the central square for a peaceful demonstration, which quickly turned into a violent riot and resulted in the death of five civilians. The incident sent a clear signal to the political leaders that in order to conduct a free and fair election in conformity with the law and reestablish the popular trust in the representative democracy, the election system needed to be changed.

Among the challenges to achieving progress in this matter, is the use of modern technologies such as electronic vote-counting machines, fingerprint recognition, live monitoring of the polling stations with video cameras in political elections in Mongolia. Authorities believed that by properly implementing these technologies, they will make the elections more efficient, cost effective as well as help eliminating various rigging incidents that have occurred in the past. Furthermore, it was believed that this would also contribute to restore public trust in the electoral process as a whole. However, there was a fear that adopting an electronic voting system would result in greater security risks and public mistrust, if the system was implemented without careful planning and preparation time. In order to analyze the effectiveness of the electoral reform in Mongolia, the author examines the following hypotheses:

- Overall benefits of the e-voting outweigh its shortcomings when compared to the traditional paper voting system in Mongolia
- E-voting will restore public trust in electoral system and its management as well as political parties of Mongolia
- E-voting will help eliminating various rigging incidents and election fraud that have occurred in past elections
- E-voting makes the voting process easier for both the citizens and the administration
- A mixed system will help solving the problems of the FPTP system such as wasted votes and disproportionality
- Mixed system along with the 20% quota for women candidates will increase women representation
- Introduction of e-voting requires time, planning and preparation to be successful

This chapter aims to analyze the electoral reform in Mongolia, the incidents that led to the change of election system and introduction of electronic voting system and their effectiveness in case of Mongolia. The political events leading to the reform of the election system will be examined in detail. The paper finds out that the use of electronic technologies in election along with the mixed system brought positive consequences in the case of Mongolia as the country struggles to prevent the negative impacts of "Dutch Disease" (increase in exploitation of natural resources and a decline in the manufacturing sector) through good governance, transparent public service and resource management.

## BACKGROUND

With a land area of 1.56 sq.km, Mongolia has the lowest population density in the world (Kohn, 2008, p. 48). However, it has the potential to become the next Asian tiger as it is going through a period of remarkable economic growth. The Mongolian economy grew by 17.3 percent in 2011 and became one of the world's fastest growing economies. The economy is projected to grow by 19% in 2013 (*Main directives for Mongolia's social and economic development in 2013*) and at an estimated 15-20% rate a year till 2020.

This strong growth is mostly attributed to foreign direct investment in mining sector. The Mongolian economy is traditionally based on mobile livestock breeding and agriculture. Mongolia has vast untapped mineral deposits of copper, coal, gold and molybdenum and is believed to possess an estimate of 1 trillion us dollars' worth of minerals in thousands of mining sites across the country. When the two major mining sources Oyu Tolgoi and Tavan Tolgoi begin operations, the economic growth is projected to accelerate even further. However, the so-called "Dutch Disease" effect, indicates that it is difficult to convert the wealth generated by the mining sector into long-term economic and human development. Dependence on the abundant natural resources tend to negatively affect the country's development.

With more than 39% of its population in poverty (UNDP, 2011), Mongolia is at a crossroad, where transparent and fair distribution of wealth may create economic prosperity, political stability and a wealthy middle class. However, if Mongolia cannot properly manage and avoid unjust wealth concentration, a widening gap between poor and rich may become even broader and deeper (Rossabi, 2005, p. 248) and Mongolia can become another resource-cursed Nigeria (Sala-i-Martin, Artadi, & Subramanian, 2003). Scholars attempted to explain thepersistence of poverty in some developing countries based on the open and limited access orders theory (North, Wallis, & Weingast, 2009; North, 2007).

The approach suggests that the attempts to implement socio-economic and political institutions of the western countries in the developing countries often fail because of the lack of understanding about how these developing societies actually behave and differ fundamentally from the developed world. "*Limited access order* creates limits on access to valuable political and economic functions as a way to generate rents" and includes almost all developing countries (North, 2007, p. 3). "*Open access order* relies on competition, open access to organizations and the rule of law to hold the society together" (North, 2007, p. 3) and represents most of the western democracies.

Western development prescriptions initially failed in Mongolia because they tried to implement institutions of the open access order directly into a limited access order. When trying to implement the policy prescriptions such as privatization, trade liberalization and decentralization imposed by the donors as a condition for financial assistance, Mongolia faced a serious economic crisis early 1990s. The "GDP per capita contracted by 34 percent between 1989 and 1992" (Bruun & Odgaard, 2013, p. 108) leaving the majority of the population in poverty, which lasted for decades. Noble Prize winner Joseph E. Stiglitz explains that one-size-fits-all policy prescriptions can worsen the situation rather than help developing countries with specific social and economic institutions (Stiglitz, 2003).

For two decades Mongolia has consolidated democracy and implemented market-supporting institutions. However, with persistent poverty, unemployment, and widespread corruption problems, the long-term prospects of Mongolian democracy is still precarious (Landman, Larizza, & McEvoy, 2006, pp. 3-4). Although Mongolia seems to satisfy the basic characteristics of an open access order (North, 2007, p. 17) and has all the formal institutions such as legislature, executive bodies, courts and elections, these institutions have informal limitations, which arise from bureaucracy and corruption at all levels of the government.

The former electoral system in Mongolia, for instance, in practice was an "incentive compatible" system (Diaz-Cayeros, Magaloni, & Weingast, 2003, p. 3) where citizens have powerful incentives to maintain the governing party in power. Distribution of cash was not only done by individual candidates, but also the two major parties made cash payout promises in the election of 2008 (Isakova, Plekhanov, & Zettelmeyer, 2012). Vote-buying practices still exist, especially in rural areas with less employment opportunities.

Even though patronage is formally prohibited, and politics and civil service are separated under the law, it is still the major issue that influences election results (Tsedev & Richard, 2005, p. 6). It is a well-known fact in Mongolia that unless a civil servant is somehow affiliated with the ruling party he or she will likely lose their job. On the other hand, supporters or relatives of the party know that they will be rewarded with administrative positions in civil service if their party wins the elections (Tsedev & Richard, 2005). In most rural areas, unemployment and poverty are widespread and not many work opportunities outside the civil service except herding animals exist. Therefore, reward of positions in public service is a matter of livelihood. Politicians use this condition to win elections and gain power (Tsedev & Richard, 2005, p. 6).

An important parliamentary election took place in Mongolia in June 2012. The election is relevant in the sense that it will be held at a time when Mongolia faces an opportunity to strengthen its economy as the mining industry booms. Experiences elsewhere show that corruption and bad governance can lead a resource rich country to become a resource cursed country (Cabrales & Hauk, 2011; Sala-i-Martin et al., 2003; Van der Ploeg, 2011).

Various studies on corruption suggest that corruption in Mongolia is widespread and pervasive.(Asia Foundation, 2010; *Transparency International Corruption Perceptions Index 2012*, 2012, *UN Convention Against Corruption Civil Society Review Mongolia 2011*, 2011; USAID, 2005; Verena, n.d.) Mongolia ranked at 94 out of 176 countries according to Transparency International Corruption Perceptions Index in 2012. (*Transparency International Corruption Perceptions Index 2012*, 2012) Public opinion surveys indicate that 53% of the people feel their government's efforts to fight corruption are ineffective and 28% feel neither effective nor ineffective ("Transparency International - Country Profiles," n.d.). Mongolia was listed as the second most corrupted country in the world in a recent study (Transparency International, 2013).

Since the transition a to market economy in the early 1990s, the country's wealth concentrated in the hands of a few people and one third of the population still lives in poverty. (Fredrick & Bernard, 2004, pp. 15–25) As in most limited access orders (North, 2007), Mongolian political leaders divided up control of the economy, limited access to valuable resources and used their privileged access for corruption. Therefore, combating widespread corruption and distribution of the wealth generated from the exploitation of the country's rich mining resources were the main focus of the 2012 electoral campaign ("Mining wealth distribution key issue in Mongolia polls," n.d., "Mongolia votes amid mining boom," 2012). After three years of prolonged debate, political parties came to an agreement on the revision of the electoral law and on the introduction of advanced technologies in the electoral process in order to restore the public trust lost in previous elections ("Automated election system law passed," 2011).

For a country of 2.7 million people, scattered around the vast, landlocked territory between Russia and China, it is interesting to see whether and how the introduction of an electronic voting system could affect the establishment of effective, transparent and responsive public service, especially when the majority of the population have no internet access. However, despite various problems, the popular demand has dramatically accelerated the use of ICT in delivery of government services ("July 1 riot was a sign for changing the election law," 2010).

## The 2008 Riot and Its Consequences

On July 1, 2008 a riot erupted after the parliamentary election. Thousands of people gathered in front of the Mongolian People's Revolutionary Party (hereinafter, MPRP) building on July 1st, 2008 and expressed their disagreement with the election results and claimed that the election was affected by fraud. The demonstration started peacefully, but ended in clashes between the police and the rioters, which resulted in a four day state of emergency, the complete burnout of the MPRP building, death of five and detention of over 700 civilians (Gregory, Christopher, & David, 2008). On July 2nd, the President of Mongolia declared a state of emergency that lasted for four days (Gregory et al., 2008).

The so-called July 1st incident brought back to evidence the issues related to a defective election system, corruption and distributive justice. There have been street demonstrations and protests almost every spring in the streets of Ulaanbaatar since the collapse of the socialism in the early 1990's. However, almost all these protests including the democratic revolution of 1990 were relatively peaceful compared to the July 1st incident, as no blood was spilled and no large-scale destruction occurred (Gregory et al., 2008).

There were different assumptions about what caused and who was responsible for the riot. Over 700 protesters including 2 leaders of the Great Party O. Magnai and J. Batzandan were arrested and about 200 of them where sentenced. The head of the Republican Party B. Jargalsaikhan was arrested for supporting the incident and the Democratic Party chief Ts. Elbegdorj stepped down. As major international human rights organizations and domestic NGO's (Amnesty International, 2010; Mongolian National Green Movement, July coalition, & Emergency Human rights monitoring and protection coalition, n.d.) condemned the human rights violations in the riots, the case was brought to justice and four chiefs of the police and security departments were convicted and sentenced for 5 civilian deaths. Following the shock of the riot there was a radical change in the attitude toward the electoral system, which was stagnant for over a decade. The main political parties came to an agreement to change the first past the post system of election and amend the legislation, which barely managed to pass through the parliament just before the next parliamentary election. These events were the turning point for election law reform in the country.

## ELECTORAL REFORM (1992-2016)

### Legal Framework

The constitution of Mongolia provides for fundamental principles of election and protects citizens' right to elect and to be elected (*Constitution of Mongolia*, 1992, art. 16.9.). "Citizens take park in the conduct of State affairs directly or through representative bodies"(*Constitution of Mongolia*, 1992, art 16.9.). Citizens enjoy right to elect from the age of eighteen and those who have reached the age of twenty-five years can be elected to the State Great Khural (*Constitution of Mongolia*, 1992, 21.3.). The constitution also stipulates principles of universal, free, direct suffrage by secret ballot (*Constitution of Mongolia*, 1992, 21.2.). Detailed electoral procedures were left for other laws leaving room for change. Since the constitution enshrines only the fundamental principles of the election, e-voting may be regulated by the legislations pertaining to election.

After three years of deliberation, the State Great Khural, adopted a revision of the Law on Election of the State Great Khural on December 15, 2011 barely six months before the next parliamentary election. Election law stipulates that no amendments or change shall be made to the election law within six months prior to the election. Discussion of the revised bill stalled in State Great Khural for three years as the major political parties had disagreement over the new changes (*Law on Election of the State Great Khural of Mongolia*, 2011). The revised law introduced many new futures to the election law. The major amendment was the change of the electoral system from majority system to a mixed system. Switch from majority system to a proportional system or a mixed electoral system has been the most frequent tendency in electoral reform around the world, especially in new democracies in Africa and former Soviet Union (Reynolds, Reilly, & Ellis, 2005, p. 23).

Other major changes to the revised law are: citizens living abroad can cast votes for the first time; 20% women quota has been introduced; automated vote counting machines; fingerprint recognition for voter registration; electoral committee and polling staff are comprised of civil servants; smart electronic identification cards for voter identification; and voter registration is shifted from the Central Election Commission to the State Registration Authority.

### Advantages and Disadvantages of the FPTP System

The country first used Block Vote system in 1992 and later abandoned it in favor of the First Past the Post (FPTP) system. In FPTP systems, voters choose only one candidate and the candidate who obtained the most votes is elected while in Block Vote systems voters can have as many votes as the available seats for that electoral district. After 1992, the Block Vote system was abolished because voters tend to give all their votes to the same party, thus producing disproportional results. In 1992 election, the Mongolian People's Revolutionary Party won 70 of the 76 seats (93% of the total seats) with over 50 percent of the votes (Jeffries, 2007, p. 14).

Compared to Block Vote system and other variations of the majority system, FPTP system is much simpler. Voters choose only one candidate and whoever receives the most vote wins. It is even possible to win by one more vote than their closest competitor (Pippa, 1997, p. 3). FPTP system has numerous advantages such as each geographical districts are equally represented; it usually strengthens two major parties and single-party government rather than a coalition government; it tends to give incentive to strong oppositions; each elected member is accountable for a certain geographical area and it is easy to understand and use (Reynolds et al., 2005, pp. 35–37). Disadvantages of the FPTP systems include difficulty to create appropriate boundaries of electoral districts; vote-splitting, results large number of wasted votes; women are less likely to be elected; and finally it causes disproportional representation.

By far the most controversial problems were the number of wasted votes and disproportional effects of FPTP system. In 2000 parliamentary election, the Mongolian People's Revolutionary Party won 51.6% of the votes and won 72 seats while the opposition, the Democratic Union only won 1 seat with 13.4 of the total votes. This disproportional pattern of votes and representation repeated number of times in parliamentary elections in last two decades and eventually led to a violent riot in 2008.

Many of these disadvantages are somewhat addressed in the newly adopted mixed system, which is a combination of majority and proportional representation systems. Mixed systems combine the benefits of both majority and proportional systems. The new system contains advantages of the proportional system such as fewer wasted votes and more inclusive representation.

## Controversies Around the Mixed System

There are two main variations of mixed system. In Mixed Member Proportional systems (MMP) proportional representation is used to compensate for disproportionality that the majority system may produce whereas in Parallel systems both majority and proportional systems coexist independently (Reynolds et al., 2005, p. 90). Initially, Mongolian variation of the mixed system was somewhat between these two models. It was not a pure Parallel system as the candidates who lost in the majoritarian district could be elected through the party list proportional representation system. At the same time, it was not MMP system because it was not designed to compensate the disproportionality effect of the majority system. The two systems run independently of each other except for the possibility of a candidate to win a seat through the party list PR system even if he lost in the majoritarian district.

This caused heated debate among scholars. A complaint was submitted to the Constitutional Court of Mongolia claiming that some provisions of the revised election law violated the equal right provision of the Constitution ("Citizens complain about the election law," 2012). According the to the law, 28 of the State Great Khural's 76 members are elected by party list proportional system while the remaining 48 members are elected by majority system. The law also gave the candidates an opportunity to be elected in the parliament through proportional system even if they had lost in the majoritarian district. A similar case was brought to the Supreme Court of Japan, but the court deemed the issue was a political matter. Contrary to that, the Mongolian Constitutional Court struck down articles 4.9; 49.1.5; 49.1.5a; 49.1.5b; 49.1.5.c; 49.1.6 of the Law on Election of the State Great Khural ("Decision 3 of the Constitutional Tsets of Mongolia," 2012). The Constitutional Tsets found that these provisions of the election law violated articles 1.2; 3.1; 14.1; 16.9; and 21.2 of the Constitution. Court decision is based on the right to elect and to be elected; general, direct, free and fair election principles; and the principle of equality before law. The Constitutional Tsets ruled that giving the candidates who lost in the majoritarian districts another chance creates an advantage for some candidates and thus is in violation of the principle of equality before law ("Decision 3 of the Constitutional Tsets of Mongolia," 2012).

As a consequence, the current electoral system is a parallel system where majority and proportional systems operate independently from each other. Although the parallel system does not guarantee overall proportionality, it results in fair degree of proportionality and helps to lessen the disproportionality problems caused by the majority system. Under the newly adopted parallel system in Mongolia, however, the disproportionality is greatly reduced. In 2012 election, the Democratic Party won 31 seats with 35.32% of the votes, Mongolian People's Party won 25 seats with 31.31% of the votes, Justice Coalition (the coalition of the newly established Mongolian People's Revolutionary Party and Mongolian National Democratic Party) won 11 seats with 22.31& of the votes and the Civil Will-Green Party won 2 seats with 5.51% of the votes.

## Gender Equality

The next important amendment was the quota for women candidates. According to Inter-Parliamentary Union's study, the proportional system is more likely to have more women elected in the parliament. Article 27.1 of the Law on Election of the State Great Khural provides for 20 percent quota for woman candidates (*Law on Election of the State Great Khural of Mongolia*, 2011, art. 27.1). In 2012 election, 11 of the 174 female candidates were elected in the parliament. Only 4 were elected via the proportional system as many women were listed lower in the list. Together they constitute 14% of the parliament members, which is lower than the world average for single house parliaments, but still higher than the 6.6% of the earlier election of 2008 ("Gender Equality in Elected Office in Asia Pacific," 2012). Thus, the new electoral system along with the 20% quota helped women candidates to gain more seats than before.

## Citizens Living Abroad Cast Votes

Mongolians living abroad, for the first time, were able to enjoy their right to vote in the election. They can cast their votes at one of the 39 diplomatic missions of Mongolia 15 days before the polling day (*Law on Election of the State Great Khural ofMongolia*, 2011, art 6.7). Nevertheless, there are several restrictions that hampered this effort. Citizens residing in a foreign country can only participate in the election by voting for a party or a coalition. This means they can only participate in electing 28 members of the parliament. They are also required to be at the embassy or the consulate office in person to register and participate in the election. Those living farther than the embassy are less likely to go and participate in the election. Thus, in the 2012 election, fewer than 3000 out of 35000 citizens living abroad were able to vote ("Mongolians residing in foreign countries vote for Democratic Party," 2012). Postal voting or internet voting will bring higher turnout. However, internet voting requires a careful consideration and preparation as constitutional issues such as secrecy of the vote may arise.

## Voter Turnout

2012 election allowed for the first-time civil society groups to observe the election process rather than the party nominees. Despite the efforts to gain back voters trust and make the election process more transparent, only 65% of the voters turned out to vote. Compared to 74.3% voter turnout in 2008, 82.2% in 2004, and 82.4% in 2000 it is record low.

Lower voter turnout cannot be attributed to the introduction of e-voting system as voter turnout was steadily declining since the first democratic election took place in 1992, where almost 96% of the voters participated. However, one possible reason could be the increased mistrust in the political 12ampaigning and electoral procedures fueled by the 2008 uprising.

## Pre-Election Controversies

On April 13, 2012, the former president Nambar Enkhbayar was arrested on corruption charges. The timing of the arrest raised questions about the purpose of the arrest. The former president Nambar Enkhbayar had established a new party and was preparing for the upcoming parliamentary election in June. The arrest came after the president refused to cooperate with the interrogators. Supporters of the Mongolian People's Revolutionary Party rallied in the street demanding release of the former president. Loyalists to Mr. N. Enkhbayar accused the arrest as politically motivated.

When the former Mongolian People's Revolutionary Party reverted to its name used prior to 1924, former president Enkhbayar split the party and established the new Mongolian People's Revolutionary Party. Mr. Enkhbayar stopped eating and drinking after his detention was extended (Amnesty International, 2012). After he refused to drink any water for 10 days, he was granted bail. All the dramas surrounding the event positively influenced the electoral success of his party. Mr. N. Enkhbayar himself was sentenced 4 years imprisonment, which was later reduced to 2.6 years of imprisonment ("Former president N.Enkhbayars sentence revised," 2012). However, the former president was pardoned later by the current president Tsakhia Elbegdorj (Kohn, 2013).

# CONSTITUTIONAL AMENDMENTS AND ELECTORAL REFORM (2015-2019)

## Election Law of 2015

In 2015, a new Election Law was adopted. Unlike previous election laws, it regulated not only the election of the State Great Khural, but also the presidential election and the election of the Citizens` Representative`s Khural at the local level. It was amended 4 times in 2016 and twice in 2017. On the positive side, the chapter 15 of the law addresses issues concerning the petition and complaints and the dispute settlement process. However, on the other hand, the article that prohibited amendment to the electoral law six months before the election was deleted. This made it possible to amend the Election Law even just before the election takes place.

Initially the election of the State Great Khural was to be held under the mixed electoral system. Forty-seven members are elected by plural system and twenty-six members under proportional electoral system. The law-makers chose a single constituency closed party list system (Law on Election, 2015, art 120.2, 3, 4.), the same system used in 2012 parliamentary election. What is different from the earlier law is that it requires parties to get one seat under the plural system in order to obtain a seat under the proportional system. Along with the 5 percent threshold, this was a restriction intentionally imposed on the new or small parties. However, the proportional system was suspended on April 22, 2016 and subsequently removed from the law on May 5, 2016 just before the parliamentary election based on the decision of the Constitutional Tsets, which found it unconstitutional.

## Constitutionality of the Proportional Electoral System

Two citizens raised the issue of constitutionality of the proportional electoral system. On April 22, 2016, the Constitutional Tsets struck down the proportional electoral system as unconstitutional. First, the Tsets found that the proportional electoral system violates the principle of direct election enshrined in the Constitution (Constitution of Mongolia, 1992, art. 21.2). The Tsets interpreted the direct election as the right of the voters to directly cast ballots for the candidate they choose without being represented by anyone, any organization and municipalities. The Tsets stated that citizens of Mongolia have the right to directly cast ballots to elect each individual member of the State Great Khural and voting for parties under the closed party list system violates this principle. However, professor B. Chimid, who is known to be the drafter of the 1992 Constitution, stated that the principle of direct election does not preclude the use of proportional electoral system (B. Chimid, *Songuulid Suraltsakhui*, 2008, p. 55).

The Tsets`s second justification for the decision is based on the minutes of the State Small Khural`s session of October 15, 1991. In 1991, Mongolia had bicameral parliament consisting of State Great Khural the upper chamber and the State Small Khural, lower chamber. Both Khurals convened sessions to adopt the Constitution of Mongolia. On that day, the State Small Khural, the lower chamber of the parliament of the time, debated on the electoral system. A mixed electoral system, two thirds are elected under majoritarian system and the rest of the members are elected under proportional system, was put to vote in the State Small Khural, but failed to get support of the majority. On this ground, the Tsets decided that the Constitution of Mongolia did not allow the proportional electoral system. This presents a logic that every constitutional issue that was discussed and failed to secure the support of the majority of the People`s Great Khural or the State Small Khural is unconstitutional.

The Tsets further tied this idea to article 21.2 and 21.3 of the Constitution, which stipulates that "members of the State Great Khural shall be... elected", and "citizen shall be elected... as a member of the State Great Khural" respectively. The Tsets came to the conclusion that these provisions dictated the direct voting for a candidate as a representative of the people rather than voting for a political party. In other words, Tsets concluded that the proportional representation system forces citizens to vote for parties rather than individual candidates, thereby violates the principle of direct voting (Decision #5 of the Constitutional Tsets of Mongolia, 2016.). In the earlier decision in 2012, the Tsets did not find the proportionality electoral system unconstitutional (Decision 3 of Constitutional Tsets of Mongolia, 2012).

The current constitutional provision stipulates that the rules of procedure for the elections of members to the State Great Khural to be prescribed by law (Constitution of Mongolia, 1992, art. 21.4.). And it has left the choice of electoral system to the discretion of the State Great Khural. However, Tsets's decision restricted the power of the State Great Khural to choose the electoral system. Because of the Tsets`s decision the Law on Election was amended on May 5, 2016 (Law on Election, 2015, art. 120.2.) and the parliamentary election of 2016 was held under majoritarian system in single member constituencies. The gender quota of 30 percent was reduced to 20 and citizens living abroad no longer enjoy the right to vote. Subsequently, the Mongolian People`s Party won 65 of the 76 parliamentary seats in the election. The governing Democratic Party lost in the election retaining only 9 seats as the result of Tsets's decision on constitutionality of the proportional system. The decision cemented the two-party system that was that dominated politics in Mongolia in the past 27 years.

## Constitutional Amendments of 2019 and Electoral Reform

Tsets`s decision resulted in strong criticism among the constitutional scholars and law communities in Mongolia. Some scholars criticized the Tsets`s historical interpretation of the Constitutional provisions (O. Munkhsaikhan, The Constitution did not prohibit the Proportional Electoral System, 2016). Even members of the State Great Khural see the decision ungrounded and 62 (out of 76) members of the State Great Khural submitted a draft of the constitutional amendment, which included a provision that will empower the State Great Khural to choose the electoral system. The proposed draft of June 5, 2018 states that "rules of procedure and the electoral system for the elections of members to the State Great Khural to be prescribed by law". The intention was to restore the State Great Khural`s power to choose the electoral system and reinstate the proportional electoral system.

The draft amendment also prohibited the amendment to or adoption of the Law on Election of the State Great Khural within one year prior to the regular elections. In 2016 parliamentary election, major changes were introduced to the Law on Election in less than two months before the election without much of a consultation with the general public. This change is necessary because any changes to the electoral system needs to facilitate the confidence of the citizens in the electoral system and the stability of the constitutionalism. Things got to a sudden turn when the President of Mongolia officially submitted his version of the constitutional amendments just before the second deliberation of the constitutional amendments submitted by the

62 members. According to the Law on Constitutional amendment, the President`s draft amendments will be considered not as separate amendments, but as a proposal to the draft amendments submitted by the State Great Khural. President`s proposal had numerous changes to the electoral system.

First the President suggested to increase the number of members of the State Great Khural from 76 to 108. Half of those members will be elected from single member constituencies whereas the other half will be elected using the proportional electoral system effectively restoring the mixed system used in the 2012 election. Next he proposed to lengthen the terms of office for members of the State Great Khural from 5 to 6 years (Draft amendment submitted by the President, art 21.2) and claimed that this would give enough time for the government to implement its policy. The last proposed constitutional change regarding election would prohibit any attempt to seize State power by influencing citizens` right to vote by illegal means.

The President`s draft required a consensus building not only between leading parties, but also with the President. The President headed conciliatory working group, which accepted many of the President`s proposals. However, all of the President`s proposals concerning the electoral system were discarded including the mixed electoral system during the second parliamentary debate. The State Great Khural even rejected the proposal submitted by the 62 parliamentary members that sought to reintroduce the proportional electoral system in the Constitution. The State Great Khural then issued a resolution to hold a referendum on the Constitutional amendments and the dissatisfied President vetoed the resolution. The State Great Khural accepted the veto and proceeded with the third deliberation of the Constitutional amendments, in which most of the proposed electoral reforms were rejected. The Constitutional amendments that were adopted in November, 2019 failed to address major issues regarding the electoral legal framework, except the prohibition on the amendment to or adoption of the Law on Election of the State Great Khural within one year prior to the regular elections, which will be effective by May, 2020.

## 2019 Law on Election of the State Great Khural

The Election Law of 2015 regulated the parliamentary, presidential as well as local elections. Different interpretations of the law in different elections resulted in disputes among stakeholders. Thus, the State Great Khural passed the Law on the Election of the State Great Khural on November 20, 2020. The initial draft of the law provided for the use of both single-member constituencies and the block vote systems each allocated with 50 and 26 seats in the 2020 parliamentary election. However, during the parliamentary deliberations, citing the lack of popular support, the parliamentary members decided to move to the block vote system, which was used in 1992 and 2008 parliamentary elections. The use of the block vote system led to the only election related violence in the country after the 2008 election, which led to the introduction of the mixed system in 2012 parliamentary election. With the proportional electoral system declared unconstitutional, the next election will be held under block vote system. The main opposition, the Democratic party seem to support this choice.

2008 election, held under block vote system, resulted in a riot and the President declared emergency. However, some attribute this partly to the manual counting of ballots that was the practice in 2008 election. In some polling stations, the manual counting of ballots took seventy-two hours. With this work load and efficiency the reliability of the election results was disputed (Report submitted by NGO Forum after the 2008 election. https://lib.ohchr.org/HRBodies/UPR/Documents/session9/MN/WSP_WomenforSocialProgressMovement.pdf). With electronic voting and vote counting commenced since 2012 election there is a hope that block vote system would not result in similar outcome.

A contentious provision of the 2019 Law on Election of the State Great Khural is that it prohibits anyone to be a candidate if the person is previously proven guilty of corruption by the court. Critics claim the provision is unconstitutional as this was intended to prevent specific persons running for election such as the former President, who has been found guilty by the court.

## ELECTRONIC VOTING SYSTEM

Electronic voting refers to the use of information and communication technologies in elections. Basically, it could mean internet voting or e-voting in uncontrolled environments (Jan & Urs, 2009; Norwegian Ministry of Local Government and Regional Development, 2006) and e-voting in polling stations or e-voting in a controlled environment. Electronic voting in polling stations is used in many countries around the world while the former is only used in few advanced democracies (IDEA, 2011, p. 6). Remote internet voting is believed to raise security considerations under certain circumstances (Jan & Urs, 2009; Rubin, 2001).

Mongolia chose to use the e-voting in polling stations in the 2012 parliamentary election primarily to restore lost trust in electoral management and increase the efficiency of the election process. Authorities believed that e-voting can help reduce election fraud and vote rigging, speed up the voter registration and vote counting process and reduce the cost of elections in the long run. Apart from known weaknesses of e-voting such as lack of transparency, potential security risks and of manipulation by insiders or by hackers, bringing and using the advanced technologies in remote areas where there is a lack of electricity and communication coverage was a major challenge for the General Election Commission ("An insider's perspective," n.d.). For the purpose of the paper, e-voting refers to e-voting in polling stations.

## Preparations

After 2008 riot, political parties agreed to reform the electoral system, which was criticized for facilitating election fraud, vote rigging and vote buying. Other stakeholders such as NGO's and international organizations were supportive of the reform. However, the two biggest parties did not reach consensus about the new election system until December 2011, about six months prior to the election. Many believed that 6 months was enough time for the technical implementation of the e-voting system. However, that was not the case in Mongolia.

In order to eliminate or minimize alleged irregularities of the voter identification and registration process such as multiple voter identification cards and fake national identification cards (Asia Foundation, n.d.), the General Election Commission introduced an electronic identification card and fingerprint recognition technology. The electronic ID cards were planned to be issued to citizens by May 2012 (*Mongolia's National eID Initiative*, 2012) before the election. However, due to time constraints the new ID cards were not used in the 2012 election.

This indicates that allowing enough time for preparation for e-voting is essential as time is required for planning, purchasing and implementing the technology and to raise citizens' awareness (Caarls & Council of Europe, 2010, pp. 21–40). Gradual implementation of e-voting system allowing enough time for preparation explains e-voting success in Brazil. Studies on the usability and feasibility of the e-voting system were carried out starting from 1986 in Brazil. This was followed by capacity building, development of hardware and software, testing and the full e-voting roll-out was implemented in 2002 election (Reynolds et al., 2005, p. 20). Brazil has a large territory and remote and sparsely populated areas similar to Mongolia. The Brazilian example shows that allowing time to build public trust and preparation can lead to successful e-voting system implementation.

In order to address the alleged problem of voter registration in 2008 parliamentary election of Mongolia, the task of maintaining voter list was transferred to the Government Authority for State Registration from the General Election Commission. The voter list was better than 2008 despite some complaints. Major issue was the home address registration. Many citizens do not register their new address when they move to a new location, which generates many names in the former address. Despite these minor issues, voter registration and preparation for the election was improved.

## Casting and Counting Votes

There have been allegations after each election about votes being rigged and how manual counting process allowed for election fraud. In 2008, manual counting lasted 72 hours on average and a number of allegations about increasing or depressing the votes of certain candidates surfaced and lasted for months without being settled. Lack of transparency was a major cause for election disputes and mistrust in electoral conduct in Mongolia.

In order to address these issues, Mongolia used optical scan tabulator machines from Dominion Voting Systems. Mr. N. Luvsanjav, Chairman of the General Election Commission said that Imagecast system from Dominion Voting is a simple to use, secure voting machine that can read one page in 5-8 seconds. "Election fraud cannot happen as the system retains a secure digital image of every ballot cast in the election… Since each machine is equipped with an eight-hour capacity battery there is no need to worry about electricity cuts" ("Сонгуулийн ерөнхий хорооны дарга Н.Лувсанжав," 2012). This was an important factor for choosing the voting machines as 30% of the polling locations, mainly in remote rural areas, operated without electricity. The machine was considered reliable as it is used over 20 states of the USA. GEC bought 2500 machines for 24 billion Mongolian tugrigs or about 17 million US dollars ("Сонгуулийн ерөнхий хорооны дарга Н.Лувсанжав," 2012). A Mongolian company "Interactive" and Information Center of the GEC is in charge of the security of the source code.

To increase transparency, video cameras were installed in polling stations. Polling station staff were civil servants rather than party-nominated partisans. In addition to that, random sample of two electoral districts in each aimag were manually counted in order to increase trust in the voting machines. Although a few discrepancies were revealed in the manual counting, overall accuracy of the automated counting machines was established.

Electronic voting systems are believed to speed up the counting of ballots and prevent election fraud that could occur during traditional system. "Newer generations of electronic voting machines appear to reduce common voter mistakes" (Alvarez, 2008, p. 25). Number of countries such as India and Brazil started using electronic voting machines while many developed countries not yet using the system amid concern for security (José Rodrigues-Filho, 2006, pp. 85–94). "Electronic voting systems can be judged on several bases, including how well its design provides for security, accuracy, ease of use, and efficiency, as well as its cost" ("Elections: Electronic Voting Offers Opportunities and Presents Challenges," 2004).

Election results came out faster and general public widely accepted the election results. Voter registration and vote casting were much more transparent and the election produced more women PMs. Around 3.6% of the votes were invalid, which was lower than the previous election. Despite the restricted time period, the election of 2012 was smooth and successful. E-voting system had an overall positive impact on conducting fair and free election in 2012 Mongolian parliamentary election.

Efficiency, accuracy and security of the electronic voting system and its effectiveness in restoring public trust was established in 2012 parliamentary election of Mongolia. E-voting in Mongolia avoided many of the problems associated with the traditional voting system and entails promising opportunity to restore voter trust in upcoming elections in Mongolia. The 2012 parliamentary election result was expected to be announced soon after the polls closing. Despite the use of the electronic voting machines, in few districts the election result was delayed because of technical problems. Nine political parties including the ruling Mongolian People's Party demanded manual hand counting in every electoral constituency and claimed that the electronic voting system was not conformity with the Constitution of Mongolia. However, they did not come up with a detail about how the e-voting violates the constitution. They blamed the automated voting machines for their failure in the election and demanded manual counting of votes. The main opposition, Democratic Party backed up the electronic voting system. When the manual count confirmed the machine counted results all parties all accepted the election results. Following the success of e-voting in the parliamentary election, 2013 presidential elections also used the e-voting system with considerable success.

## Confidence in the Voting Machines

In 2016 parliamentary election, electronic voting and counting machines were used. However, there was a lack of confidence in the machines among voters. Due to this, a manual recount in half of the polling stations around the country was introduced in the law. The law requires the manual recount will be held randomly and shall be held within 24 hours.[1] Despite distrust in the electronic voting system, the effectiveness and trustworthiness automated vote counting system become undeniable as it was used in the elections of 2012 and 2016.

## CONCLUSION

Public confidence in the electoral system diminished due to the allegation of election fraud and the July 1, 2008 riot. The electronic voting system, along with the parallel electoral system, offered a number of benefits such as increased proportionality, higher female representation in the parliament, and efficient elections. The most significant benefit of the e-voting system was the restoration of popular trust in the electoral system. However, trust in the political parties was not restored as demonstrated by the decline in voter turnout. Despite limited time to implement e-voting, the parliamentary election of 2012 was remarkably well-organized and peaceful. Allegations of vote rigging and election fraud, which were a hallmark of previous elections, did not surface in connection with the e-voting in the 2012 election. Some parties demanded a manual recount of the ballots, but later accepted the election result. The positive outcomes of the 2012 election can be attributed to the introduction of e-voting and the electoral reform. Thus, the overall benefits of e-voting outweigh the potential weaknesses of the system when compared to the traditional paper voting in Mongolia.

Many of the achievements of 2012 election were reversed by the 2015 Election law. The woman quota shrank, the proportional electoral system was declared unconstitutional, overseas voters can no longer vote in the parliamentary election. The 2019 constitutional amendments offered solutions to many electoral issues, but none of them except one was supported by the ¾ majority in the State Great Khural. Despite many setbacks and suspicions and distrust, the electronic voting system restored public confidence in the democratic election, increased the efficiency and reliability of the vote counting in 2012 and 2016 elections, and contributed to the development of constitutionalism in Mongolia.

## REFERENCES

Alvarez, R. M. (2008). *Electronic elections the perils and promises of digital democracy*. Princeton, NJ: Princeton University Press. Retrieved from http://site.ebrary.com/id/10367266

*Amnesty International.* (2010). Mongolia Submission to the UN Human Rights Committee. Author.

Amnesty International. (2012). *Mongolian Authorities must respect the human rights of former Mongolian president following his arrest.* Author.

An insider's perspective. (n.d.). *International IDEA*. Retrieved August 25, 2013, from http://www.idea.int/elections/an-insiders-perspective-working_with_elections_in_mongolia.cfm

Asia Foundation. (2010). *Mongolian Corruption Benchmarking Survey: Tenth Semi-Annual Corruption Benchmarking Survey in Mongolia*. Author.

Asia Foundation. (n.d.). Voter Registration in Mongolia. *Assessment of Policies and Practices*.

Automated election system law passed. (2011, November 10). *Daily News*.

Bruun, O., & Odgaard, O. (2013). *Mongolia in Transition: Old Patterns, New Challenges*. Routledge. doi:10.4324/9780203036648

Caarls, S., & Council of Europe. (2010). *E-voting handbook: key steps in the implementation of e-enabled elections*. Strasbourg: Council of Europe Pub.

Cabrales, A., & Hauk, E. (2011). The Quality of Political Institutions and the Curse of Natural Resources*. *Economic Journal (London), 121*(551), 58–88. doi:10.1111/j.1468-0297.2010.02390.x

Chimid, B. (2008). *Songuulid Suraltsakhui*, 55.

Citizens complain about the election law. (2012). Retrieved August 24, 2013, from http://english.news.mn/content/97971.shtml

Constitution of Mongolia (1992).

Decision of the Constitutional Tsets of Mongolia. (2012, March 28). Retrieved August 24, 2013, from http://conscourt.gov.mn/index.php/mn/decisions/conclusion-pack/324-2012-03-28

Diaz-Cayeros, A., Magaloni, B., & Weingast, B. R. (2003). Tragic brilliance: Equilibrium hegemony and democratization in Mexico. *Hoover Institution, Stanford University*. Retrieved from http://www.bsos.umd.edu/gvpt/calvo/articulos/diaz-cayero-magaloni-weingast-tragicbrilliance.pdf

Elections: Electronic Voting Offers Opportunities and Presents Challenges. (2004). Retrieved August 25, 2013, from http://www.gao.gov/products/GAO-04-975T

Former president N. Enkhbayars sentence revised. (2012). Retrieved August 25, 2013, from http://english.news.mn/content/127921.shtml

Fredrick, N., & Bernard, W. (2004). *Privatization, Income Distribution and Poverty: the Mongolian Experience*. UNDP.

Gender Equality in Elected Office in Asia Pacific. (2012). UNDP.

Gregory, D., Christopher, K., & David, S. (2008). *The end of post-socialism? An account of the 1st of July riots in Ulaanbaatar*. Mongolia: Academic Press.

IDEA. (2011). *Introducing Electronic Voting*. IDEA.

Isakova, A., Plekhanov, A., & Zettelmeyer, J. (2012). *Managing Mongolia's resource boom*. Retrieved from http://www.ebrd.com/downloads/research/economics/workingpapers/wp0138.pdf

Jan, G., & Urs, G. (2009). *Three case studies from Switzerland: E-Voting*. Academic Press.

Jeffries, I. (2007). *A Guide to Economic and Political Developments* Taylor & Francis. doi:10.4324/9780203099667

José Rodrigues-Filho, C. A. (2006). *E-Voting in Brazil - The Risks to Democracy*. Academic Press.

July 1 riot was a sign for changing the election law. (2010). Retrieved August 24, 2013, from http://politics.news.mn/content/5795.shtml

Kohn, M. (2008). Mongolia. Lonely Planet.

Kohn, M. (2013). Former Mongolian President Pardoned After 2012 Graft Conviction. *Bloomberg*. Retrieved August 25, 2013, from http://www.bloomberg.com/news/2013-08-01/former-mongolian-president-pardoned-after-2012-graft-conviction.html

Landman, T., Larizza, M., & McEvoy, C. (2006). *State of democracy in Mongolia: A desk study*. UNDP.

Law on Election of the State Great Khural of Mongolia (2011).

Mining wealth distribution key issue in Mongolia polls. (n.d.). *GlobalPost*. Retrieved August 24, 2013, from http://www.globalpost.com/dispatch/news/afp/130623/mining-wealth-distribution-key-issue-mongolia-polls

Mongolia votes amid mining boom. (2012, June 28). *BBC*. Retrieved from http://www.bbc.co.uk/news/world-asia-18621555

Mongolia's National eID Initiative (2012).

Mongolian National Green Movement, July Coalition, & Emergency - Human Rights Monitoring and Protection Coalition. (n.d.). *July 1 and human rights abuse*. Author.

Mongolians residing in foreign countries vote for Democratic Party. (2012). Retrieved August 24, 2013, from http://www.infomongolia.com/ct/ci/4413

Munkhsaikhan, O. (2016). *The Constitution did not prohibit the Proportional Electoral System*. http://www.trends.mn/n/5588?fbclid=IwAR04wLs9wrJdvTgiBFABpcSfFohXI77csTYPPgVjRTyRHxRJ1

North, D. C. (2007). *Limited access orders in the developing world: A new approach to the problems of development* (Vol. 4359). World Bank. doi:10.1596/1813-9450-4359

North, D. C., Wallis, J. J., & Weingast, B. Rt. (2009). Violence and the Rise of Open-Access Orders. *Journal of Democracy*, *20*(1), 55–68. doi:10.1353/jod.0.0060

Norwegian Ministry of Local Government and Regional Development. (2006). *Electronic voting - challenges and opportunities*. Author.

Pippa, N. (1997). Choosing Electoral Systems: Proportional, Majoritarian and Mixed Systems. *International Political Science Review*, *18*(3), 297–312. doi:10.1177/019251297018003005

Report submitted by NGO Forum after the 2008 election. (n.d.). Retrieved from https://lib.ohchr.org/HRBodies/UPR/Documents/session9/MN/WSP_WomenforSocialProgressMovement.pdf

Reynolds, A., Reilly, B., & Ellis, A. (2005). *Electoral system design: the new international IDEA handbook*. International Institute for Democracy and Electoral Assistance.

Rossabi, M. (2005). *Modern Mongolia: from khans to commissars to capitalists*. University of California Press.

Rubin, A. (2001). *Security considerations for remote electronic voting over the Internet*. Retrieved from http://books.google.com/books?hl=en&lr=&id=qPpdyIycDfMC&oi=fnd&pg=PA105&dq=%22to+improve+our+elections+is%22+%22the+danger+that+outside+of+a+public+polling+place,+a+voter+could%22+%22of+our+carefully+controlled+elections+as+we+know+them.+The+relevant+issues+are%22+&ots=GK97OmejPm&sig=5_JG5uPIZYrsW3EhRoH84_G_xrQ

Sala-i-Martin, X., Artadi, E. V., & Subramanian, A. (2003). *How can Nigeria Address the Natural Resource Curse*. IMF Working Paper 03/129. Washington, DC: International Monetary Fund.

Stiglitz, J. E. (2003). *Globalization and Its Discontents*. W.W. Norton.

Transparency International. (2013). *Global Corruption Barometer*. Author.

Transparency International. (2012). *Corruption Perceptions Index 2012*. Author.

Tsag. (2012). *N. Luvsanajav, Chief of General Election Commission: No One Can Intentionally Confuse US Counting Machine*. Retrieved August 25, 2013, from http://www.24tsag.mn/content/7192.shtml

Tsedev, D., & Richard, P. (2005). *Institutional Change in Mongolia: Balancing Waves of Reform*. ADB.

UN Convention Against Corruption Civil Society Review Mongolia. (2011). UN.

USAID. (2005). Assessment of Corruption in Mongolia: Final Report (No. DFD-100-03-00139-00). Ulaanbaatar: USAID.

Van der Ploeg, F. (2011). Natural resources: Curse or blessing? *Journal of Economic Literature*, *49*(2), 366–420. doi:10.1257/jel.49.2.366

Verena, F. (n.d.). *Democratization and Corruption in Mongolia*. Academic Press.

## ENDNOTE

1 Law on Elections, 2015, art 97.13.

# Compilation of References


Abbott, J., MacDonald, A., & Givens, J. (2013). New Social Media and (Electronic) Democratization in East and Southeast Asia: Malaysia and China Compared. *Taiwan Journal of Democracy*, *9*(2), 105–137.

Abramo, C. W. (2010). Percepções pantanosas: a dificuldade de medir a corrupção. *Revista Novos Estudos, 34*(3).

Activists in China's Wukan village being "spied upon, harassed". (2012, March 10). *BBC Monitoring Asia Pacific-Political.*

Ai, W. (2018) Ai Weiwei: The artwork that made me the most dangerous person in China. *The Guardian.* Retrieved from https://www.theguardian.com/artanddesign/2018/feb/15/ai-weiwei-remembering-sichuan-earthquake

Alexander, A. (2011, February 9). Internet Role in Egypt's Protests. *BBC News*. Retrieved from: http://www.bbc.co.uk/news/world-middle-east-12400319

Álvarez, I. (2012). *El frente en defensa Wirikuta Tmatshima Wahaa: un acercamiento sistémico a los Nuevos Movimientos Sociales* [Wirikuta Tmatshima Wahaa's Defense Front: A Systemic Approach to New Social Movements] (Unpublished Graduate Dissertation). National Autonomous University of Mexico, Mexico City, Mexico.

Álvarez, I. (2019). Wirikuta en disputa. Reivindicaci6n del patrimonio ejidal y minero en el Altiplano Potosino [Wirikuta in dispute. Claim of the ejido and mining heritage in the Potosian Plateau]. *Sociedad y Ambiente Society and Environment*, *7*(19), 31–54.

Alvarez, R. M. (2008). *Electronic elections the perils and promises of digital democracy*. Princeton, NJ: Princeton University Press. Retrieved from http://site.ebrary.com/id/10367266

Amnesty International. (2009) Justice Denied: Harassment of Sichuan earthquake survivors and activists. *Amnesty International Publications*. Retrieved from https://www.amnesty.org/en/documents/ASA17/018/2009/en/

*Amnesty International*. (2010). Mongolia Submission to the UN Human Rights Committee. Author.

Amnesty International. (2012). *Mongolian Authorities must respect the human rights of former Mongolian president following his arrest.* Author.

An insider's perspective. (n.d.). *International IDEA*. Retrieved August 25, 2013, from http://www.idea.int/elections/an-insiders-perspective-working_with_elections_in_mongolia.cfm

Anderson, C. J., & Tverdova, Y. (2003). Corruption, Political Allegiances, and Attitudes Toward Government in Contemporary Democracies. *American Journal of Political Science*, *47*(1), 91–109. doi:10.1111/1540-5907.00007

Ansell, C. K. (2001). Legitimacy: Political. In *International Encyclopedia of the Social & Behavioral Sciences* (pp. 8704–8706). New York: Elsevier. doi:10.1016/B0-08-043076-7/01167-0

Anti-Japan protests continue in China over territorial row. *BBC Monitoring Asia Pacific-Political.*

Areddy, J. (2011, December 29). China Blasts High-Speed Rail System. *The Wall Street Journal.* Retrieved from: http://online.wsj.com/article/SB10001424052970204632204577126121683353312.html

Arsu, S. (2014, March 20). Turkish Officials Block Twitter in Leak Inquiry. *The New York Times.* Retrieved from http://www.nytimes.com

Arthur, C. (2011, January 28). Egypt Cuts Off Internet Access. *The Guardian.* Retrieved from: http://www.guardian.co.uk/technology/2011/jan/28/egypt-cuts-off-internet-access

Asia Foundation. (2010). *Mongolian Corruption Benchmarking Survey: Tenth Semi-Annual Corruption Benchmarking Survey in Mongolia.* Author.

Asia Foundation. (n.d.). Voter Registration in Mongolia. *Assessment of Policies and Practices.*

Asia: A new third rail. (2011, July 30). *The Economist.* Retrieved from: www.economist.com/node/21524914

Automated election system law passed. (2011, November 10). *Daily News.*

Bac, M. (2001). Corruption, Connections and Transparency: Does a Better Screen Imply a Better Scene? *Public Choice*, *107*(1-2), 87–96. doi:10.1023/A:1010349907813

Bandurski, D. (2012, January 30). The Lagacy of Wukan. *China Media Project.* Retrieved from: http://cmp.hku.hk/2012/01/30/18237/

Bandursky, D. (2009). Looking Back on Chinese Media Reporting of School Collapses. *China Media Project.* Retrieved from http://chinamediaproject.org/2009/05/07/one-year-of-chinese-media-reporting-on-school-collapses/

*Bantam Books, Inc. v. Sullivan*, 373 U.S. 58, 70 (1963).

Basil, W. J. (2004). Democratization: Transitions and Conflicts in Asia and the Pacific. *Asian Human Rights Commission.* Retrieved June 30, 2009 from: http://www.ahrchk.net/ pub/mainfile.php/pandp/164/

Bay Area Rapid Transit. (2011, December 1). *Extraordinary circumstances only for cell phone interruptions.* Retrieved from https://www.bart.gov/news/articles/2011/news20111201

Beetham, D. (1993). In Defence of Legitimacy. *Political Studies*, *41*(3), 488–491. doi:10.1111/j.1467-9248.1993.tb01651.x

Berggruen, N., & Gardels, N. (2013). *Intelligent Governance for the 21st Century—A Middle Way between West and East.* Polity Press.

Bernard, E., Steen-Johnsen, K., & Wollebaek, D. (2012, January), *How do social media change the conditions for civic and political mobilization*? Paper presented at 3rd. International Conference on Democracy as Idea and Practice, University of Oslo.

Bertot, J. C., Jaeger, P. T., & Hansen, D. (2012). The Impact of Polices on Government Social Media Usage: Issues, Challenges, and Recommendations. *Government Information Quarterly*, *29*, 30–40. doi:.giq.2011.04.004 doi:10.1016/j

Bertot, J. C., Jaeger, P. T., & Grimes, J. M. (2010). Using ICTs to Create a Culture of Transparency: e-Government and Social Media as Openness and Anti-Corruption Tools for Societies. *Government Information Quarterly*, *27*(3), 264–271. doi:10.1016/j.giq.2010.03.001

Blight, G., Rininsland, A., Rogers, S., & Torpey, P. (2012, April 16). How Open is Your Internet? An interactive map. *The Guardian*. Retrieved at: http://www.guardian.co.uk/technology/datablog/interactive/2012/apr/16/web-filtering-censorship

*Bobby Bland, Daniel Ray Carter, JR., David W. Dixon, Robert W. McCoy, John C. Sandhofer, and Debra H. Woodard v. B.J. Roberts*, 2012, 857 F. Supp. 2d 599 (E.D. Va).

Bradsher, K., & Fackler, M. (in press). Protesters across China vent anger at Japan; Dispute over island chain adds fuel to tensions as nationalism is inflamed. *The International Herald Tribune.*

Breslin, S. (2008). *Democratising One-Party Rule? Political reform, national, and legitimacy in the People's Republic of China* (Working Paper no. 67). Retrieved from FRIDE website: www.fride.org/download/wp67_china_one_party_eng2_sep08.pdf

Breuer, A. (2012). *The Role of Social Media in Mobilizing Political Protest. Evidence from the Tunisian Revolution*. Discussion Paper 10/2012, German Development Institute.

Brink, D. (2010). *Web Security in the Cloud: More Secure! Compliant! Less Expensive!* Retrieved from http://www.aberdeen.com/aberdeen-library/6572/RA-web-cloud-security-compliance.aspx

Brinkerhoff, D. W., & Goldsmith, A. A. (2002). *Clientelism, Patrimonialism and Democratic Governance: An Overview and Framework for Assessment and Programming*. Retrieved August 08, 2007 from http://www.abtassoci- ates.com/reports/2002601089183_30950.pdf

Bristow, M. (2009, November 23). Prison term for Chinese activist. *BBC News*. Retrieved from http://news.bbc.co.uk/go/pr/fr/-/2/hi/asia-pacific/8373573.stm

British Broadcasting Corporation. (2011, January 28). *Egypt severs internet connection amid growing unrest.* Retrieved from https://www.bbc.com/news/technology-12306041

British Columbia. (2011). *Citizen satisfaction with provincial government services, 2010/2011 citizen consultation on service delivery*. Data presented by Darin Thomson of the Ministry of Justice, British Columbia, at the 2012 American Bar Association Dispute Resolution Section Conference, Washington, DC.

British Columbia. (2012). *Civil resolution tribunal act*. Retrieved from http://www.leg.bc.ca/39th4th/3rd_read/gov44-3.htm

British Columbia. (n.d.). *Introduction to online service delivery*. Retrieved from http://www2.gov.bc.ca/assets/gov/topic/D9B6F4A42E7F6EF9641DDB53AF6D8B10/intro_to_osd/introduction_to_online_service_delivery.pdf

Bruun, O., & Odgaard, O. (2013). *Mongolia in Transition: Old Patterns, New Challenges*. Routledge. doi:10.4324/9780203036648

Buckley, C. (2008, June 16). China prosecurtors join quake school collapse probe. *Reuters*. Retrieved from https://www.reuters.com/article/idINIndia-34078920080616

Bulwa, D. Wyatt Buchanan and Matthew Yi (2009, January 15), *Behind murder charge against ex-BART officer*. SFGate, Retrieved from http://www.sfgate.com/bayarea/article/Behind-murder-charge-against-ex-BART-officer-3254683.php)

Caarls, S., & Council of Europe. (2010). *E-voting handbook: key steps in the implementation of e-enabled elections*. Strasbourg: Council of Europe Pub.

Cabrales, A., & Hauk, E. (2011). The Quality of Political Institutions and the Curse of Natural Resources*. *Economic Journal (London)*, *121*(551), 58–88. doi:10.1111/j.1468-0297.2010.02390.x

Cardoso, F. H. (2005, Sept.). Political Parties. *Foreign Policy*, 41.

Carey, J., & Shugart, M. (Eds.). (1998). *Executive Decree Authority*. Cambridge, UK: Cambridge University Press; doi:10.1017/ CBO9780511609305

Checks and Balances. (2008). In *Encyclopædia Britannica*. Retrieved July 03, 2008 from: http://www.britannica.com/Ebchecked/ topic/108283/checks-and-balances

Chen, Y. (2013). *Public Opinion Guidance of Wechat Dissemination in Major Events*. Retrieved from http://www.cqvip.com/qk/81149x/201413/661594554.html

Chimid, B. (2008). *Songuulid Suraltsakhui*, 55.

China feels the quake online. (2008, May 16). *BBC News*. Retrieved from http://news.bbc.co.uk/2/hi/asia-pacific/7402612.stm

China Internet Network Information Center. (2012, January). *Statistical Report on the Internet Development in China* (Abridged Edition). Retrieved at: http://www1.cnnic.cn/IDR/ReportDownloads/201209/P020120904421720687608.pdf

China sacks railway spokesman over remarks on train crash. (n.d.). *BBC Monitoring Asia Pacific-Political.*

China Struggles to Tame Microblogging Masses. (2011, September 8). *The Independent.*

China: 5,335 students killed or missing after 2008 quake. (2009, May 10). *CNN*. Retrieved from http://www.cnn.com/2009/WORLD/asiapcf/05/07/china.quake.deaths/index.html

China: Five surrender to police for vandalism during anti-Japan protests. *BBC Monitoring Asia Pacific-Political.*

China's Territorial Disputes. (2012, October 12). *The New York Times*. Retrieved from http://topics.nytimes.com/top/news/international/countriesandterritories/china/territorial-disputes/index.html

China-Japan dispute. (2010, September 26). *The Korea Herald.*

Chinese man held over earthquake photos. (2008, July 31). *The Telegraph*. Retrieved at https://www.telegraph.co.uk/news/worldnews/asia/china/2477024/Chinese-man-held-over-earthquake-photos.html

Chinese web users condemn video of protester beating owner of Japan-made car. *BBC Monitoring Asia Pacific-Political.*

Chung, J. (2011). *Weibo and 'Iron Curtain 2.0' in China: Who Is Winning the Cat-and-Mouse Game?* East Asia Institute Issue Briefing, No. MASI 2011-07. Retrieved at: http://www.eai.or.kr/type/panelView.asp?code=eng_report&catcode=&idx=10700&bytag=p

Citizens complain about the election law. (2012). Retrieved August 24, 2013, from http://english.news.mn/content/97971.shtml

*Clark v. Community for Creative Non-Violence*, 468 U.S. 288, 104 S. Ct. 3065, 82 L. Ed. 2d 221 (1984).

Clem, W. (in press). Seeing through all the talk of transparency; An 'open, transparent' investigation in the Wenzhou train crash? We've heard that kind of talk before. *South China Morning Post.*

*Cohen v. California*, 403 U.S. 15 (1971).

Coleman, S. (2001). The Transformation of Citizenship? In B. Axford & R. Huggings (Eds.), *New Media and Politics* (pp. 110–121). London: SAGE Publications. doi:10.4135/9781446218846.n5

Collier, D., & Levitsky, S. (1997). Democracy with "Adjectives". *World Politics*, *49*(03), 430–451. doi:10.1353/wp.1997.0009

Constitution of Mongolia (1992).

Coonan, C. (2009, May 4). China's quake cover-up. *The Independent.* Retrieved from https://www.independent.co.uk/news/world/asia/chinas-quake-cover-up-1678542.html

Coonan, C. (in press). Wukan villagers end protest after getting key concessions. *The Irish Times.*

Dahl, R. (1956). *A Preface to Democratic Theory*. Chicago: The University of Chicago Press.

Dahl, R. (1971). *Polyarchy*. New Haven, CT: Yale University Press.

*Davidson v. Plowman* 715 Fed. Appx. 298 (4th Cir. 2018).

De Mauleón, H. (2012, Sept. 9). De la red a las calles [From Social Media to the Streets]. *Nexos*.

Dean, J., & Page, J. (2011, July 30). Trouble on the China Express. *The Wall Street Journal.*

Decision of the Constitutional Tsets of Mongolia. (2012, March 28). Retrieved August 24, 2013, from http://conscourt.gov.mn/index.php/mn/decisions/conclusion-pack/324-2012-03-28

Demick, B. (in press). China-Japan island dispute fuels protests; Nationalist fervor by activists leads to vandalism and efforts to reach one isle. *Los Angeles Times.*

Demick, B., & Makinen, J. (in press). Beijing's hand seen in anti-Japanese protests; Officials appeared to have guided marchers, but then some rallies went out of control. *Los Angeles Times.*

Deng, Y., & Moore, T. G. (2004). China Views Globalization: Towards a New Power Politics. *The Washington Quarterly*, *27*(3), 115–136. doi:10.1162/016366004323090296

Díaz Alba, C. (2013). Tres miradas desde el interior de #YoSoy132 [Three looks from inside #YoSoy132]. *Desacatos*, 43.

Diaz, L., Hart Dubner, B., & Parent, J. (2007). When is a "rock" an "island"? – Another unilateral declaration defies "norms" of international law. *Michigan State Journal of International Law,* 519-555.

Diaz-Cayeros, A., Magaloni, B., & Weingast, B. R. (2003). Tragic brilliance: Equilibrium hegemony and democratization in Mexico. *Hoover Institution, Stanford University*. Retrieved from http://www.bsos.umd.edu/gvpt/calvo/articulos/diaz-cayero-magaloni-weingast-tragicbrilliance.pdf

Dissent in China: Of Development and Dictators. (2011, August 6). *The Economist*. Retrieved at: http://www.economist.com/node/21525419

East Asia Territorial Disputes. Geopolitical Implications. *Business Monitor Online*. Retrieved October 29, 2012, from http://wwws.moreheadstate.edu:2052/hottopics/lnacademic/?

Eimer, D. (in press). A new Orwellian euphemism for stifling online dissent; David Eimer says 'cultural security' targets the messengers, not the issues. *South China Morning Post.*

Elections: Electronic Voting Offers Opportunities and Presents Challenges. (2004). Retrieved August 25, 2013, from http://www.gao.gov/products/GAO-04-975T

Ely, J. H. (2010). *Democracia e Desconfiança; uma Teoria do Controle Judicial de Constitucionalidade*. São Paulo: WMF Martins Fontes.

Facebook, 2013, Brief of Facebook, inc. As *amicus curiae* in support of Plaintiff-Appellant Daniel Ray Carter, Jr. And in support of *vacatur*, in the United States Court of Appeals for the Fourth Circuit, Bobby Bland, Daniel Ray Carter, Jr., David w. Dixon, Robert W. Mccoy, John C. Sandhofer, and Debra H. Woodward, Plaintiffs-Appellants, v. B.J. Roberts, Defendant-Appellee.

FindLaw Writers and Editors. (n.d.). *Civil Cases vs. Criminal Cases: Key Differences*. Retrieved from https://litigation.findlaw.com/filing-a-lawsuit/civil-cases-vs-criminal-cases-key-differences.html

Fishkin, J. S. (2015). *Quando o Povo Fala: Democracia Deliberativa e Consulta Pública*. Curitiba, Brazil: Instituto Atuação.

Five Years after a Quake, Chinese Cite Shoddy Reconstruction. (2013, May 13). *National Public Radio* retrieved from https://www.npr.org/sections/parallels/2013/05/14/183635289/Five-Years-After-A-Quake-Chinese-Cite-Shoddy-Reconstruction

Flanagan, E. (2011, July 29). China's All Powerful Rail Ministry Takes Heat for Crash. *NBC News*. Retrieved from http://behindthewall.nbcnews.com/_nv/more/section/archive?date=2011/7

Ford, P. (2010, February 9). China sentences quake activist Tan Zuoren. *Christian Science Monitor*. Retrieved from https://www.csmonitor.com/World/Asia-Pacific/2010/0209/China-sentences-quake-activist-Tan-Zuoren

Former president N. Enkhbayars sentence revised. (2012). Retrieved August 25, 2013, from http://english.news.mn/content/127921.shtml

Foucault, M. (1991). Governmentality. In G. Burchell, C. Gordon, & P. Miller (Eds.), *The Foucault Effect, Studies in Governmentality*. Chicago: The University of Chicago Press.

Fredrick, N., & Bernard, W. (2004). *Privatization, Income Distribution and Poverty: the Mongolian Experience*. UNDP.

*Freedman v. Maryland*, 380 U.S. 51 (1965).

Freedom House. (2007). *Freedom in the World*–2007. Retrieved December 06, 2007 from: http://www.freedomhouse.org/template. cfm?page=351&ana_page=333&year=2007

Gainous, J., Wagner, K., & Abbott, J. (2015). Civic Disobedience: Does Internet Use Stimulate Political Unrest in East-Asia? *Journal of Information Technology & Politics*, *12*(2), 219–236. doi:10.1080/19331681.2015.1034909

Gender Equality in Elected Office in Asia Pacific. (2012). UNDP.

Gladwell, M. (2010, Oct. 4). Small Change: Why the Revolution Will not be Tweeted. *The New Yorker*.

González S., Borge-Holthoefer, J, Rivero, A., & Moreno, Y. (2011). The Dynamics of Protest Recruitment through an Online Network, *Scientific Reports*, *10*(1038), 1-7.

González, J. L. (2012). #YoSoy132: Participación política 2.0 en México [#YoSoy132: Political participation 2.0 in Mexico]. *Diálogo Político*, *29*(3), 77–104.

Google Glass. (2014, April 6). Retrieved from http://www.google.com/glass/start/

Gregory, D., Christopher, K., & David, S. (2008). *The end of post-socialism? An account of the 1st of July riots in Ulaanbaatar*. Mongolia: Academic Press.

Griggs, B. (2013, March 10). Beyond the DMZ: Tweeting from North Korea. *CNN*. Retrieved at: http://www.cnn.com/2013/03/10/tech/social-media/social-media-north-korea/index.html

Grindle, M. S. (2007). Good Enough Governance Revisited. *Development Policy Review*, *25*(5), 553–574. doi:10.1111/j.1467-7679.2007.00385.x

Gross, D. (2012, December 3). Syria Caused Internet Blackout Security Firm Says. *CNN*. Retrieved at: http://www.cnn.com/2012/11/30/tech/web/syria-internet/index.html

Hamilton, A. J. J., & Madison, J. (1787-88). *The Federalist Papers*. Retrieved February 20, 2014 from: http://www.gutenberg.org/ etext/18

Hidy, K. (2019). Social Media Use and Viewpoint Discrimination: A First Amendment Judicial Tightrope Walk with Rights and Risks Hanging in the Balance. *Marq. L. Rev.*, *102*, 1045-1085.

Hoffman, A. (2005). Political Parties, Elector- al Systems and Democracy: A Cross-national Analysis. *European Journal of Political Research*, *44*(2), 231–242. doi:10.1111/j.1475-6765.2005.00225.x

Holbig, H., & Gilley, B. (2010). Reclaiming Legitimacy in China. *Politics & Policy*, *38*(3), 395–422. doi:10.1111/j.1747-1346.2010.00241.x

Howard, Agarwal, & Hussain. (2011, October). *The Dictators' Digital Dilemma: When Do States Disconnect Their Digital Networks?* Brookings. www.brookings.edu/papers/2011/10_dictators_digital_network.aspx

Hughes, C. (2000). Nationalsim in Chinese Cyberspace. *Cambridge Review of International Affairs*, *13*(2), 195–209. doi:10.1080/09557570008400309

Huntington, S. (1991). *The Third Wave: Democratization in the Late Twentieth Century*. Norman, OK: University of Oklahoma Press.

IAAO. (2010). *Standard on property tax policy*. Retrieved from http://docs.iaao.org/media/standards/Standard_on_Property_Tax_Policy.pdf

IAAO. (2013). *Standard on assessment appeal*. Retrieved from http://www.iaao.org/uploads/Assessment_Appeal_2013.pdf

IDEA. (2011). *Introducing Electronic Voting*. IDEA.

In China anti-Japan protests over disputed islands get some zing. (2012, Sept. 13). *L.A. Times*.

Information Office of the State Council of China. (2008, July 6). *Wenchuan Earthquake has Killed 69,196 People and Left 18,379 Missing*. Retrieved from http://news.sina.com.cn/c/2008-07-06/162615881691.shtml

International Monetary Fund. (1997). *Good Governance: The IMF's Role*. Retrieved January 30, 2007 from: http://www.imf.org/ external/pubs/ft/exrp/gvern/govindex.htm

International Telecommunications Union. (n.d.). http://www.itu.int/en/ITU-D/Statistics/Pages/stat/default.aspx visited 3/25/2014

Internet World Stats. (n.d.). *World Internet Usage and Population Statistics June 30, 2012*. Retrieved on January 20, 2014 from: http://www.internetworldstats.com/stats.htm

Isakova, A., Plekhanov, A., & Zettelmeyer, J. (2012). *Managing Mongolia's resource boom*. Retrieved from http://www.ebrd.com/downloads/research/economics/workingpapers/wp0138.pdf

Jan, G., & Urs, G. (2009). *Three case studies from Switzerland: E-Voting*. Academic Press.

Jeffries, I. (2007). *A Guide to Economic and Political Developments* Taylor & Francis. doi:10.4324/9780203099667

Jiao, P. (in press). Party chief rant lights debate; One city official has brought on praise and mockery over remarks to villagers protesting against land seizures. *South China Morning Post.*

Jiao, P. (in press). Weibo puts the heat on officials over tragedy; Despite propaganda to downplay a high-speed train crash, microblog site and media force authorities to give more accurate information on the accident. *South China Morning Post.*

Jing, M., Zhou, Y., & Ma, D. (2014). *The Communication Mode, Characteristics and Reflection of WeChat.* Retrieved from https://wenku.baidu.com/view/054a486fe55c3b3567ec102de2bd960590c6d91a.html

José Rodrigues-Filho, C. A. (2006). *E-Voting in Brazil - The Risks to Democracy*. Academic Press.

July 1 riot was a sign for changing the election law. (2010). Retrieved August 24, 2013, from http://politics.news.mn/content/5795.shtml

Kaufman, S. (2008, May 20). *China's Government Praised for Easing Media Restrictions*. Retrieved from https://web.archive.org/web/20080703063114/http://www.america.gov/st/freepress-english/2008/May/20080520153034esnamfuak0.7039911.html

Kaufmann, D., Kraay, A., & Zoido-Lobaton, P. (1999). Aggregating Governance Indicators. *World Bank Policy Research Working Paper No. 2195*. Retrieved November 15, 2007 from: http://www-wds.worldbank.org/ex- ternal/default/WDSContentServer/IW3P/IB/1999/10/23/000094946_99101105050593/additional/115515322_20041117135531.pdf

Kersbergen Kees, V., & Waarden, V. (2004). 'Governance' as a Bridge Between Disciplines: Cross-disciplinary Inspiration Regarding Shifts in Governance and Problems of Governability, Accountability and Legitimacy. *European Journal of Political Research*, *43*(2), 143–171. doi:10.1111/j.1475-6765.2004.00149.x

Khamis, S., & Vaughn, K. (2011). Cyberactivism in the Egyptian Revolution: how civic engagement and citizen journalism tilted the balance. *Arab Media & Society*, 14. Retrieved at: http://www.arabmediasociety.com/?article=769

King, G., Pan, J., & Roberts, M. (In press). How Censorship in China Allows Government Criticism but Silences Collective Expression. *American Political Science Review*. Retrieved at: http://gking.harvard.edu/files/gking/files/censored.pdf

King, L. A. (2003). Deliberation, Legitimacy and Multilateral Democracy. *Governance: An International Journal of Policy, Administration and Institutions, 16*(1), 23–50. doi:10.1111/1468-0491.t01-1-00203

Kleinerman, B. A. (2007). Can the Prince Really be Tamed? Executive Prerogative, Popular Apathy, and the Constitutional Frame in Locke's Second Treatise. *The American Political Science Review, 101*(2), 209–222. doi:10.1017/S000305540707013X

*Knight First Amendment Institute at Columbia University v. Trump*, 928 F.3d 226 (2nd Cir. 2019).

Kohn, M. (2013). Former Mongolian President Pardoned After 2012 Graft Conviction. *Bloomberg*. Retrieved August 25, 2013, from http://www.bloomberg.com/news/2013-08-01/former-mongolian-president-pardoned-after-2012-graft-conviction.html

Kohn, M. (2008). Mongolia. Lonely Planet.

Krishnan, A. (2011). A Unique Protest in China. *The Hindu*. Retrieved at: http://www.thehindu.com/news/international/a-unique-protest-in-china/article2715292.ece

Kuhn, A. (2008, May 4). Chinese seek answers in student quake deaths. *National Public Radio*. Retrieved at https://www.npr.org/templates/story/story.php?storyId=103727282

Kwong, P. (2009, May 11). Silencing the Messenger. *New York Times*. Retrieved from https://www.nytimes.com/2009/05/12/opinion/12iht-edkwong.html?searchResultPosition=10

Lafont, C. (2012). Agreement and Consent in Kant and Habermas—Can Kantian Constructivism be Fruitful for Democratic Theory? *The Philosophical Forum, 43*(3), 277–295. doi:10.1111/j.1467-9191.2012.00425.x

LaFraniere, S. (in press). A risky about-face in China; Yielding to the merit of the Wukan protests, politician defuses revolt. *The International Herald Tribune*.

LaFraniere, S. (in press). Media Blackout in China After Wreck. *The New York Times*.

Lai, A. (2012, April 4). Chinese artist Ai Weiwei places himself under home surveillance. *CNN*. Retrieved from https://www.cnn.com/2012/04/04/world/asia/ai-weiwei-webcams/index.html

Lamberti, M. J. (2011). *"Abajo del amanecer" el corazón del universo en disputa. El caso de la instalación de empresas mineras en Wirikuta* ["Down the Dawn" the heart of the Universe in dispute. The case of the instalation of mining companies in Wirikuta] (PhD dissertation). El Colegio de México.

Landman, T., Larizza, M., & McEvoy, C. (2006). *State of democracy in Mongolia: A desk study.* UNDP.

Lao, M. (2018, May 10). A Decade-long search for justice on student deaths. *Inkstone*. Retrieved at https://www.inkstonenews.com/china/parents-call-investigation-shoddy-school-buildings-sichuan-earthquake-anniversary/article/2145722

Larru, J. (2009). Corrupción y ayuda al desarrollo. Evidencias, teoría y aplicaciones para España. *Cuadernos Económicos de ICE, 78*, 261–280. doi:10.32796/cice.2009.78.5976

Latinobarómetro. (2006). *Latinobarometro Report: Online Databank—2006.* Retrieved December 03, 2007 from: http://latinobaro- metro.org/index.php?id=101

Lau, M. (2011, December 22). Olive branch brings life back to normal in Wukan; Villages remove banners and cancel protest after party officials offer to free detainees and back a committee in dispute over land grabs and death in custody. *South China Morning Post.*

Lau, M. (2012, January 10). Victory over corruption; A protest by villagers over illegal land sales forced the normally hardline mainland authorities to negotiate – and could set a precedent in the future. *South China Morning Post.*

Lau, M. (2012, October 6) Pressure on protest leader turned party chief. *South China Morning Post.*

Lau, M. (2012, October 6). Wukan finds democracy hard going; Guangdong villagers freely elected their leaders after months of protest, but running their own affairs and getting their land back is proving tough. *South China Morning Post.*

Law on Election of the State Great Khural of Mongolia (2011).

Leonard, J. (2010, July 8). Former BART officer convicted of involuntary manslaughter. *The LA Times*. Retrieved from http://articles.latimes.com/2010/jul/08/local/la-me-bart-verdict-20100709

Levin, D. (2013, May 11). *Social Media in China Fuel Citizen Response to Quake*. Retrieved from https://www.nytimes.com/2013/05/12/world/asia/quake-response.html?login=email&auth=

Li, L., & O'Brein, K. (2010). Protest Leadership in Rural China. In P. Hays & S. Rosen (Eds.), *Chinese Politics* (pp. 85–108). London, UK: Routledge.

Linz, J. J., & Stepan, A. (1996). *Problems of Democratic Transition and Consolidation: Southern Europe, South America, and Post-Communist Europe*. Baltimore, MD: Johns Hopkins.

Liu, S., & Chang, T. (2016). One Disaster, Three Institutional Responses: Legitimation crisis and competing discourses in China. *Journalism Studies*, *9*(1), 1–23.

Locke, J. (1988). *Two Treatises of Government.* New York: Cambridge University Press; doi:10.1017/CBO9780511810268

MacKinnon, R. (2011). China's "Networked Authoritarianism". *Journal of Democracy*, *22*(2), 32–46. doi:10.1353/jod.2011.0033

Magistad, M. (2012). How Weibo is Changing China. *Yale Global Online Magazine*. Retrieved at: http://yaleglobal.yale.edu/content/how-weibo-changing-china

Makinen, J. (in press) Isle dispute fuels offbeat protests by Chinese; Japanese vehicles and online porn are being dragged into Beijing's tug of war with Tokyo. *Los Angeles Times.*

Makinen, J. (in press). Anti-Japan rallies break out in more China cities; Violence continues as protesters target diplomatic posts and business in a dispute over uninhabited isles. *Los Angeles Times.*

Manjiao, C. (2011). The Unhelpfulness of Treat Law in Solving the Sino-Japan Sovereign Dispute over the Diaoyu Islands. *East Asia Law Review,* 163-189.

Many media outlets defy China gag on Wenzhou train crash reporting. (n.d.). *BBC Monitoring Asia Pacific-Political.*

Marshall, M., & Jaggers, K. (2005). *Political Regime Characteristics and Transitions, 1800-2004— Polity IV Project Dataset User's Manual.* Retrieved December 07, 2007 from: http://www.cidcm.umd.edu/polity/data/pol- reg.htm

McCoy, T. (2014, March 21). Turkey bans Twitter — and Twitter explodes. *The Washington Post.* Retrieved from http://www.washingtonpost.com/news/morning-mix/wp/2014/03/21/turkey-bans-twitter-and-twitter-explodes/?tid=pm_pop

*McIntyre v. Ohio Elections Com'n*, 514 U.S. 334, 115 S. Ct. 1511, 131 L. Ed. 2d 426 (1995).

Miller, G., O'Keefe, E., & Goldman, A. (2014, March 11). Feinstein: CIA searched Intelligence Committee computers. *Washington Post.* Retrieved from http://www.washingtonpost.com/world/national-security/feinstein-cia-searched-intelligence-committee-computers/2014/03/11/982cbc2c-a923-11e3-8599-ce7295b6851c_story.html

Mill, J. S. (1989). *On Liberty*. London: The Walter Scott Publishing Co.

Mining wealth distribution key issue in Mongolia polls. (n.d.). *GlobalPost*. Retrieved August 24, 2013, from http://www.globalpost.com/dispatch/news/afp/130623/mining-wealth-distribution-key-issue-mongolia-polls

Mong, A., & Gu, B. (2011, July 25). Deadly Crash in China Provokes Outrage. *NBC News.* Retrieved from http://behindthewall.nbcnews.com/_nv/more/section/archive?date=2011/7

Mongolia votes amid mining boom. (2012, June 28). *BBC*. Retrieved from http://www.bbc.co.uk/news/world-asia-18621555

Mongolian National Green Movement, July Coalition, & Emergency - Human Rights Monitoring and Protection Coalition. (n.d.). *July 1 and human rights abuse*. Author.

Mongolians residing in foreign countries vote for Democratic Party. (2012). Retrieved August 24, 2013, from http://www.infomongolia.com/ct/ci/4413

Mongolia's National eID Initiative (2012).

Moore, M. (2008). China earthquake brings out citizen journalists. *The Telegraph*. Retrieved from https://www.telegraph.co.uk/news/worldnews/asia/china/1950212/China-earthquake-brings-out-citizen-journalists.html

Moore, M. (2001). Political Underdevelopment: What Causes Bad Governance? *Public Management Review*, *1*(3), 385–418. doi:10.1080/14616670110050020

Moura & Corbellini. (2019). A eleição disruptiva: por que Bolsonaro venceu. Rio de Janeiro: Record.

Munkhsaikhan, O. (2016). *The Constitution did not prohibit the Proportional Electoral System.* http://www.trends.mn/n/5588?fbclid=IwAR04wLs9wrJdvTgiBFABpcSfFohXI77csTYPPgVjRTyRHxRJ1

National Institute of Geography and Statistics. (2004). La población Indígena en México [Indigenous Population in Mexico]. Author.

National Institute of Geography and Statistics. (2005). *INEGI*. Aguascalientes, Mexico: Conteo de Poblacion y Vivienda.

National Jurist. (2009, Nov. 2). *Scare you to death, work you to death, and bore you to death.* Author.

*Neb. Press Ass'n v. Stuart*, 427 U.S. 539, 559 (1976).

Ni, Y., Kong, X., & Zhang, Y. (2011, July 27). Xinhua Insight: Relatives' sorrow amid claims and doubts regarding China Train Crash. *Xinhua General News Service*. Retrieved at: http://news.xinhuanet.com/english2010/indepth/2011-07/27/c_131013761.htm

No Justice NoB. A. R. T. (2014, April 6). http://nojusticenobart.blogspot.com/2009/01/what-is-no-justice-no-bart.html

Norris, P. (2002). *Democratic Phoenix. Reinventing político activism*. Cambridge: Cambridge University Press. doi:10.1017/CBO9780511610073

North, D. C., Wallis, J. J., Webb, S. B., & Weingast, B. R. (2007). *Limited Access Orders in the Developing World: A New Approach to the Problems of Development*. The World Bank Independent Evaluation Group. doi:10.1596/1813-9450-4359

North, D. C., Wallis, J. J., & Weingast, B. Rt. (2009). Violence and the Rise of Open-Access Orders. *Journal of Democracy*, *20*(1), 55–68. doi:10.1353/jod.0.0060

Norwegian Ministry of Local Government and Regional Development. (2006). *Electronic voting - challenges and opportunities*. Author.

O'Brien, K. (Ed.). (2008). *Popular Protest in China*. Cambridge, MA: Harvard University Press. doi:10.4159/9780674041585

O'Brien, K., & Stern, R. (2008). Introduction: Studying Contention in China. In K. O'Brien (Ed.), *Popular Protest in China* (pp. 11–25). Cambridge, MA: Harvard University Press. doi:10.4159/9780674041585-intro

O'Donnell, G. (n. d.). *Delegative Democracy?* Kellogg Institute Working Paper. Retrieved July 12, 2007 from: http://kellogg.nd.edu/ publications/workingpapers/WPS/172.pdf

O'Donnell, G. (1998). Horizontal Account- ability in New Democracies. *Journal of Democracy*, *9*(3), 112–126. doi:10.1353/jod.1998.0051

Olesen, A. (2012, August 2). Sina Weibo, China's Twitter, leads microblog craze. *The Huffington Post.* Retrieved at: http://www.huffingtonpost.com/2012/08/02/sina-weibo-chinas-twitter_n_1732300.html?vi

Olson, M. (1991). Autocracy, Democracy and Prosperity. In R. J. Zeckhause (Ed.), *Strategy and Choice*. Cambridge, MA: MIT Press.

Osnos, E. (2012, October 22). Boss Rail: The disaster that exposed the underside of the boom. *The New Yorker*. Retrieved at: http://www.newyorker.com/reporting/2012/10/22/121022fa_fact_osnos?currentPage=1

Ostrove, E. & Leber, Y. (2019, April) Silencing Critics in the Age of America's First Tweeter-In-Chief, Where Social Media Meets the First Amendment. *New Jersey Lawyer, The Magazine*, 317.

Owen, D. (2017). The New Media's Role in Politics. In *The Age of Perplexity. Rethinking The World we knew*. Madrid: BBVA.

Parry, G., & Moran, M. (Eds.). (1994). *Democracy and Democratization*. New York: Routledge; doi:10.4324/9780203198551

Parsons, T., Ji, C., & Kirby, E. (2008). Stress changes from the 2008 Wenchuan earthquake and increased hazard in the Sichuan basin. *Nature*, *454*(7203), 509–510. doi:10.1038/nature07177 PMID:18600260

Pereira, C., Power, T. J., & Rennó, L. (2005). Under What Conditions do Presidents Re- sort to Decree Power? Theory and Evidence from: The Brazilian Case. *The Journal of Politics*, *67*(1), 178–200. doi:10.1111/j.1468-2508.2005.00313.x

Perlez, J. (in press). China Accuses Japan of Stealing After Purchase of Group of Disputed Islands. *The New York Times.*

Pew Research Center's Internet & American Life Project. (2010). *Government online: The internet gives citizens new paths to government services and information.* Retrieved from http://www.pewinternet.org/~/media//Files/Reports/2010/PIP_Government_Online_2010_with_topline.pdf

Pew Research Center's Internet & American Life Project. (n.d.). *Trend data (adults).* Retrieved from http://pewinternet.org/Trend-Data-(Adults)/Whos-Online.aspx

Pildes, R. (2008). The Legal Structure of Democracy. In *Whittington, Keith; Kelemen, Daniel & Caldeira, Gregory. The Oxford Handbook of Law and Politics*. Oxford, UK: Oxford University Press. doi:10.1093/oxfordhb/9780199208425.003.0018

Pippa, N. (1997). Choosing Electoral Systems: Proportional, Majoritarian and Mixed Systems. *International Political Science Review*, *18*(3), 297–312. doi:10.1177/019251297018003005

Police break up protest by parents of China earthquake victims. (2008, June 3). Retrieved from https://www.theguardian.com/world/2008/jun/03/chinaearthquake.china?gusrc=rss&feed=net workfront

Policinski, G. (2011, August 19). BART Cell Shutdown a Landmark in Cyber-Assembly. *First Amendment Center*. http://www.firstamendmentcenter.org/bart-cell-shutdown-a-landmark-in-cyber-assembly

Polito, M. (1994). Wit, Will and Governance in Early Modern Legal Literature. *Mosaic*, *27*(4), 15–34.

Posner, R. (2012). *A Problemática da Teoria Moral e Jurídica*. São Paulo: Editora WMF Martins Fontes.

Protecting Cyberspace as a National Asset Act of 2010, S.3480 - 111th Congress. (2010).

Protests in Taiwan over Japan's purchase of disputed islands 23 September. (n.d.). *BBC Monitoring Asia Pacific-Political.*

Qiang, X. (2011a). The Battle for the Chinese Internet. *Journal of Democracy*, *22*(2), 47–61. doi:10.1353/jod.2011.0020

Qiang, X. (2011b). The Rise of Online Public Opinion and its Political Impact. In S. L. Shirk (Ed.), *Changing Media, Changing China*. Oxford, UK: Oxford University Press.

Qu, Y., Huang, C., Zhang, P., & Zhang, J. (2011). Harnessing social media in response to major disasters. *CSCW 2011 Workshop: Designing Social and Collaborative Systems for China*.

Quiang, X. (2004). A Bold new voice – Lu Yuegang's Extraordinary Open Letter to Authorities. *China Digital Times*. Retrieved on March 27, 2014 from https://chinadigitaltimes. net/2004/07/a-bold-new-voice-lu-yuegangs- extraordinary-open-letter-to-authorities/

Qu, Y., Wu, P., & Wang, X. (2009). *Online Community Response to Major Disaster: A Study of Tianya Forum in the 2008 Sichuan Earthquake*. Presented at the *42nd Hawaii International Conferences on System Sciences*, Honolulu, HI.

Rabinovich-Einy, O., & Katsh, E. (2012). Lessons from online dispute resolution for dispute systems design. In M. S. Wahab, E. Katsh, & D. Rainey (Eds.), *Online dispute resolution: Theory and practice* (pp. 39–59). The Hague, Netherlands: Eleven International Publishing.

Rainey, D., & Katsh, E. (2012). ODR and government. In M. S. Wahab, E. Katsh, & D. Rainey (Eds.), *Online dispute resolution: Theory and practice* (pp. 237–249). The Hague, Netherlands: Eleven International Publishing.

Ramos. (2015). Corrupción y ayuda al desarrollo. Evidencias, teoría y aplicaciones para España. Madrid: Universidad CEU-San Pablo.

Rapoza, K. (2011). Can China's Renren survive Sina's Weibo? *Forbes*. Retrieved at: http://www.forbes.com/sites/kenrapoza/2011/08/16/can-chinas-renren-survive-sinas-weibo/

Report submitted by NGO Forum after the 2008 election. (n.d.). Retrieved from https://lib.ohchr.org/HRBodies/UPR/Documents/session9/MN/WSP_WomenforSocialProgressMovement.pdf

Reynolds, A., Reilly, B., & Ellis, A. (2005). *Electoral system design: the new international IDEA handbook*. International Institute for Democracy and Electoral Assistance.

Rossabi, M. (2005). *Modern Mongolia: from khans to commissars to capitalists*. University of California Press.

*Roth v. U.S.*, 354 U.S. 476, 77 S. Ct. 1304, 1 L. Ed. 2D 1498 (1957).

Rousseau, J. J. (1913). *The Social Contract*. London: J.M. Dent & Sons.

Rubin, A. (2001). *Security considerations for remote electronic voting over the Internet*. Retrieved from http://books.google.com/books?hl=en&lr=&id=qPpdyIycDfMC&oi=fnd&pg=PA105&dq=%22to+improve+our+elections+is%22+%22the+danger+that+outside+of+a+public+polling+place,+a+voter+could%22+%22of+our+carefully+controlled+elections+as+we+know+them.+The+relevant+issues+are%22+&ots=GK97OmejPm&sig=5_JG5uPIZYrsW3EhRoH84_G_xrQ

Saavedra, M. (2014). Sistema de protesta: política, medios y el #YoSoy 132 [Protest system: politics, media and the #YoSoy 132]. *Sociológica*, *29*(82), 83–123.

Sala-i-Martin, X., Artadi, E. V., & Subramanian, A. (2003). *How can Nigeria Address the Natural Resource Curse*. IMF Working Paper 03/129. Washington, DC: International Monetary Fund.

Schudson, M. (1995). *The Power of News*. Cambridge, MA: Harvard University Press.

Schumpeter, J. (1950). *Capitalism, Social- ism and Democracy*. New York: Harper & Brothers.

Shakespeare, W. (1598). *Henry IV*. London: Stationers Company.

Sharp, G. (2010). *Dictatorship to Democracy—A Conceptual Framework for Liberation*. Boston: The Albert Einstein Foundation.

Sichuan earthquake killed more than 5,000 pupils, says China. (2009, May 7). *The Guardian*. Retrieved from https://www.theguardian.com/world/2009/may/07/china-quake-pupils-death-toll

Sichuan earthquake: government prohibits investigations into collapsed schools. (2009, April 24). *Asia News IT.* Retrieve from http://www.asianews.it/news-en/Sichuan-earthquake:-government-prohibits-investigations-into-collapsed-schools-15072.html

Siy, Feld, Bankston, Llansó, & Tien. (2012, April 30). *Comments Of Public Knowledge.* Center For Democracy & Technology, Electronic Frontier Foundation, Benton Foundation, Free Press, Minority Media And Telecommunications Council, National Hispanic Media Coalition, And Open Technology Institute At The New America Foundation.

Skach, C. (2005). Constitutional Origins of Dictatorship and Democracy. *Constitutional Political Economy, 16*(4), 247–368. doi:10.100710602-005-4316-x

*Snyder v. Phelps*, 131 S. Ct. 1207 (2011).

Sowell, T. (2011). *Intellectuals and Society.* New York: Basic Books.

Spruhan, P. (2015, March). Indian Law. *The Federal Lawyer*, 14–15.

Stanley, J. W., & Weare, C. (2004). The Effects of Internet Use on Political Participation: Evidence from an Agency Online Discussion Forum. *Administration & Society, 36*(5), 245–269. doi:10.1177/0095399704268503

Statistical Survey Report on the Internet Development in China, Abridged Edition. (2008, July). *China Internet Network Information Center.* Retrieved at: http://www.apira.org/data/upload/pdf/Asia-Pacific/CNNIC/22rdreport-en.pdf

Stiglitz, J. E. (2003). *Globalization and Its Discontents.* W.W. Norton.

Taiwan fishermen plan protest trip to disputed islands with Japan. (n.d.). *BBC Monitoring Asia Pacific-Political.*

Taiwan legislature demands firm stance on disputed islands. (n.d.). *BBC News Monitoring Asia Pacific-Political.*

Taylor, B. (2012). *Troubled Waters in the East China Sea.* The Australian.

Teacher, S., & Liu, S. was Released to Serve his Reeducation-Through-Labor Sentence Outside of Labor Camp. (2008, September 26). *Human Rights in China.* Retrieved at https://www.hrichina.org/en/content/220

Tergesen, A. (2007, November 4). How to reduce your property taxes. *Bloomberg BusinessWeek.* Retrieved from http://www.businessweek.com/stories/2007-11-04/how-to-reduce-your-property-taxes

Tessler, J. (2010). *Poll: Americans connecting with government online.* Retrieved from http://www.pewinternet.org/Media-Mentions/2010/AP-Americans-Connecting-With-Government-Online.aspx

The Nation. (2011). Mainland protests. *The Nation.*

The Nation. (2012). The China Post. *The Nation.*

The Net Promoter Community. (n.d.). *The net promoter score and system.* Retrieved from http://www.netpromoter.com/why-net-promoter/know/

The New York Times. (2007). *Phone Shows Ap- ple's Impact of Consumer Products.* Retrieved on December 19, 2013 from: http://www. nytimes.com/2007/01/11/technology/11cnd- apple.html?pagewanted=print&_r=0

The Subjective Value Inventory. (n.d.). *What do people value when they negotiate?* Retrieved from http://subjectivevalue.com/

Tilly, C., & Tarrow, S. (2007). *Contentious Politics.* Boulder, CO: Paradigm.

Tolbert, C. J., Mossberger, K., & McNeal, R. (2008). Institutions, policy innovation, and e-government in the American States. *Public Administration Review, 68*(3), 549–563. doi:10.1111/j.1540-6210.2008.00890.x

Transparency International. (2007). *Transparency International Corruption Perceptions Index: A Short Methodological Note.* Retrieved January 15, 2008 from: http:// chapterzone.transparency.org/content/down- load/23966/358199

Transparency International. (2012). *Corruption Perceptions Index 2012.* Author.

Transparency International. (2013). *Global Corruption Barometer.* Author.

Transparency International. (n.d.a). *Our History.* Retrieved on April 11, 2014 from http:// www.transparency.org/whoweare/history

Transparency International. (n.d.b). *Government Defence Anti-corruption Index – Methodology (Long).* Retrieved on April 11, 2014 from http://government.defenceindex.org/ sites/default/files/documents/GI-methodology-long.pdf

Trayer, T. (2012). Wukan holds local elections, displays civil disobedience. *Human Rights Brief,* 32-34.

Trindle, J. (2008, May 29). *China Questions Construction Quality.* Retrieved from https://web.archive.org/web/20080613062138/http://voanews.com/english/2008-05-29-voa12.cfm

Tsag. (2012). *N. Luvsanajav, Chief of General Election Commission: No One Can Intentionally Confuse US Counting Machine.* Retrieved August 25, 2013, from http://www.24tsag.mn/content/7192.shtml

Tsedev, D., & Richard, P. (2005). *Institutional Change in Mongolia: Balancing Waves of Reform.* ADB.

Tudou launches punchy Nasdaq IPO, tests investor bid for China. (n.d.). *EuroWeek.*

*Turner Broadcasting System, Inc. v. Federal Communications Commission,* 512 U.S. 622, 114 S. Ct. 2445, 129 L. Ed. 2d 497 (1994).

Turnsek, M., & Jankowski, N. W. (2008, April). *Social Media and Politics: Theoretical and Methodological Considerations in Designing a Study of Political Engagement.* Paper presented at Politics Web 2: An International Conference, New Political Communication Unit, Royal Holloway, University of London.

Twitter.com. (n.d.a). *New user FAQs.* https://support.twitter.com/groups/50-welcome-to-twitter/topics/203-faqs/articles/13920-new-user-faqs

Twitter.com. (n.d.b). *Help Center.* https://help.twitter.com/en/using-twitter/blocking-and-unblocking-accounts

Tze-wei, N. (in press). Microblogs' contribution to rule of law praised; Editor of official review of legal developments cites online campaign against child abduction. *South China Morning Post.*

Tze-wei, N. (in press). Payout offer raised after public outcry; Cash compensation to families who lost relatives in the rail disaster is increased from 500,000 yuan to 915,000 yuan, but some critics say this is still too low. *South China Morning Post.*

Tze-wei, N. (in press). Survival of two and a half year-old girl moves many hearts. *South China Morning Post.*

U.S. Const. amend. I.

Ukkusuri, S., Zhan, X., Sadri, A., & Ye, Q. (2014). Use of Social Media Data to Explore Crisis Informatics: Study of 2013 Oklahoma Tornado. *Transportation Research Record: Journal of the Transportation Research Board, 2459*(1), 110–118. doi:10.3141/2459-13

UN Convention Against Corruption Civil Society Review Mongolia. (2011). UN.

*United States v. Wei Sang Phua*, 100 F.Supp.3d 1040 (2015).

USAID. (2005). Assessment of Corruption in Mongolia: Final Report (No. DFD-100-03-00139-00). Ulaanbaatar: USAID.

Valenzuela, S. (2013). Unpacking the Use of Social Media for Protest Behavior: The Role of Information, Opinion Expression and Activism. *The American Behavioral Scientist, XX*(X), 1–23.

Van der Ploeg, F. (2011). Natural resources: Curse or blessing? *Journal of Economic Literature, 49*(2), 366–420. doi:10.1257/jel.49.2.366

Vega, A., & Merino, J. (2011). *Ciudadanos.mx* [Citizens.mx]. Random House Mondadori.

Veloso, G. R. (2006). *Clientelismo: Uma Instituição Política Brasileira* (MA dissertation). Universidade Nacional de Brasília. Retrieved April 04, 2008 from: http://repositorio.bce. unb.br/handle/10482/1431

Verena, F. (n.d.). *Democratization and Corruption in Mongolia.* Academic Press.

Video: Tan Zuoren's Sichuan Earthquake Investigation. (2009, December 13). Retrieved from https://chinadigitaltimes.net/2009/12/video-tan-zuorens-sichuan-earthquake-investigation/

*Wallace v. Jaffree*, 472 U.S. 38, 105 S. Ct. 2479, 86 L. Ed. 2d 29, 25 Ed. Law Rep. 39 (1985).

Ward, D. (2016). In Her Words: Recognizing and Preventing Abusive Litigation Against Survivors of Domestic Violence. Seattle Journal for Social Justice, 14(2), 11.

Weaver, M. (2008, June 3). *Police break up protest by parents of China earthquake victims.* Retrieved from https://www.theguardian.com/world/2008/jun/03/chinaearthquake.china?gusrc=rss&feed=networkfront

Weber, M. (1947). *From Max Weber: Essays in Sociology*. London: Gerth & Mills.

Wen, P. (in press). Purge lifts lid on inner workings of politburo. *Sydney Morning Herald.*

Wen, P. (in press). Wang Yang wouldn't admit it, of course, but he may be the future face of China. *Sydney Morning Herald.*

Wermiel. (2016). SCOTUS for law students: Indian cases at the Court. *SCOTUSblog*. Retrieved from http://www.scotusblog.com/2016/01/scotus-for-law-students-indian-cases-at-the-court/

Weston, T. B., & Jensen, L. M. (2012). *China In and Beyond the Headlines*. Lanham, MA: Rowman and Littlefield.

Williams, E. (2019), First Amendment Protection Against Curtailment of Access to, or Retaliation for Communications on, Social Media. *American Law Reports*, 38 A.L.R. Fed. 3d Art. 5 Sec. 5.

Winerman, L. (2009). Crisis Communication. *Nature*, *457*(7228), 376–378. doi:10.1038/457376a PMID:19158766

Wines, M., Ansfield, J., & LaFraniere, S. (2011, August 8). China's Premier Seeks Reforms and Relevance. *The New York Times*.

Wirikuta's Defense Front. (n.d.). Retrieved from http://www.frenteendefensadewirikuta.org

Wolff. (2018). Fogo e fúria: por dentro da Casa Branca de Trump. São Paulo: Objetiva.

Wong, E. (2009, August 12). Chinese Artist Says He Was Barred From Rights Advocate's Trial. *New York Times.* Retrieved from https://www.nytimes.com/2009/08/13/world/asia/13china.html

Wong, E. (in press). Demonstrators Who Took Over Chinese Village Halt Protest. *The New York Times.*

Wong, G. (2010, February 9). Chinese activist who probed collapsed schools in Sichuan quake jailed. *The Toronto Star*. Retrieved from https://www.thestar.com/news/world/2010/02/09/chinese_activist_who_probed_collapse_of_schools_in_sichuan_quake_jailed.html

World Bank. (2019). *Individuals using the Internet (% of the Population).* Retrieved from World Bank Open data: data.worldbank.org

World Bank. (n. d.). *Good Governance and Its Benefits on Economic Development: An Overview of Current Trends.* Retrieved January 30, 2007 from http://www.worldbank.org/ wbi/governance/pdf/eduardpresentation_mu- nich_inwent.pdf

Wu, J. (2008, May 12). School building quality to be probed. *China Daily*. Retrieved from http://www.chinadaily.com.cn/china/2008-05/17/content_6692329.htm

Wu, J. (2008, May 17). *School building quality to be probed.* Retrieved from http://www.chinadaily.com.cn/china/2008-05/17/content_6692329.htm

Xinhua notes active response of "we-media" to Chinese train crash. (2011, July 25). *BBC Monitoring Asia Pacific-Political.*

Xu, L. (2008, June 7). *The Power of Chinese Netizens After the Earthquake.* Retrieved from http://english.ohmynews.com/articleview/article_view.asp?menu=c10400&no=382780&rel_no=1&isPrint=print

Xu, B., & Albert, E. (2017). *Media Censorship in China.* Council on Foreign Relations.

Yan, A. (in press). Rail-crash relatives are resigned to the outcome; One year after the Wenzhou high-speed train disaster, families continue to mourn but the authorities have failed to answer their grievances. *South China Morning Post.*

Yang, G. (2008). Contention in Cyberspace. In K. O'Brien (Ed.), *Popular Protest in China* (pp. 126–143). Cambridge, MA: Harvard University Press. doi:10.4159/9780674041585-008

Yardley, J. (2008, May 25). *Chinese Are Left to Ask Why Schools Crumbled.* Retrieved from https://www.nytimes.com/2008/05/25/world/asia/25schools.html?action=click&module=RelatedCoverage&pgtype=Article®ion=Footer

Yongnian, Z. (in press). Guangdong's reform and China's future. *The Straits Times.*

York, G. (2016, December 01). *Beijing can't muzzle outrage over deadly collapsed schools.* Retrieved from https://www.theglobeandmail.com/news/world/beijing-cant-muzzle-outrage-over-deadly-collapsed-schools/article674653/

Youpee-Roll, A. (2016). *Supreme Court Makes Up Indian Law Decisions.* Retrieved from Indianz - http://www.indianz.com/News/2016/02/18/april-youpeeroll-supreme-court.asp

Yu, X. (2015). *Matthew effect, barrel effect and trickle-down effect of Wechat transmission.* Retrieved from http://www.doc88.com/p-2572805204612.html

Zhai, K. (2012, September 23). After the flames died down, some national soul-searching: Protest coverage ran from calls for rational behaviour to tough questions on patriotism. *South China Morning Post.*

Zhang, X. (2011). Mass media in China. In *Understanding Chinese Society* (pp. 156–169). New York, NY: Routledge.

Zittrain, J. L. (2008). *The Future of the Internet-- And How to Stop It.* New Haven, CT: Yale University Press & Penguin.

Zuckerberg, M. (2012). *Letter from Mark Zuckerberg.* Washington, DC: The United States Securities and Exchange Commission. Retrieved on February 26, 2014 from http://www.sec.gov/Archives/ed- gar/data/1326801/000119312512034517/ d287954ds1.htm#toc287954_10

# About the Contributors

**Sam B. Edwards III** obtained his Juris Doctor from the Northwestern School of Law of Lewis and Clark College and his LLM. in international environmental law from Nagoya University, Graduate School of Law in Nagoya, Japan. He is an associate professor of legal studies at Quinnipiac University in Connecticut, USA. Professor Edwards holds a license to practice law in California, the Territory of Guam, The Commonwealth of the Northern Mariana Islands, and Yap State in the Federated States of Micronesia. Professor Edwards' research focuses on the intersection of law, culture, and technology. Professor Edwards has practiced, studied, and taught law in the United States, Kenya, The Federated States of Micronesia, Japan, and Thailand. He has published in the areas of law and culture, international trade in endangered species, business law, environmental valuation in Micronesia, and online business transactions for small and medium enterprises.

**Diogo Santos** obtained his Bachelors of Laws in 2003 from the Federal University of Maranhão, in Brazil, Diogo Santos was granted a Japanese Ministry of Education (MEXT) scholarship to pursue a masters of law degree at the Graduate Shcool of Law in 2007 and a PhD in International Development at the Graduate School of International Development in 2010, both at Nagoya University. Since returning to Brazil, Dr. Santos has served in various academic positions, such as Director of the School of Law of the State University of Maranhão, professor of International Law, economics and Constitutional Theory at the Dom Bosco University in São Luís, Brazil, and professor of International Law in the Federal University of Maranhão, professor of Constitutional Law and Department Chair at Estacio College. Diogo has also published a number of articles in the fields of Intellectual Property Law, Environmental Law, Democracy, Legitimacy and Presidential decrees in various international journals.

* * *

**Atzimba Baltazar Macías** is a Partner of Probatio SC a consultancy firm on public policy analysis. During 2013, she was the Director of the Coordination and Monitoring Unit of the Secretary General at the Federal Institute of Access to Information (IFAI) in Mexico, where she previously served from 2002 to 2007 as a case researcher of appeals submitted to IFAI related to denials of information requests. From 2010 to 2013, Ms. Baltazar worked as the Program Officer for the Mentoring Government Capacity for Budget Transparency and Participation Program at the International Budget Partnership where she assisted in the development of training, learning, and technical materials to strengthen government capacities for budget transparency, particularly in the provision of useful information and opportunities for citizen participation. Ms. Baltazar also was a research assistant at the Center for Economic Research and Teaching in Mexico City in the Municipal Reform Project sponsored by the Ford Foundation. She holds a Master of Public Administration in Public and Nonprofit Management and Policy from the New York University and a BA in Public Administration from El Colegio de Mexico. Ms. Baltazar has also published various articles on transparency, access to information, agrarian and decentralization policies in Mexico.

**James Robert Masterson** is Assistant Dean, MPA Director, and Associate Professor of Political Science at Morehead State University. He recently served as the Director of International Education and of MSU's Department of State funded Chinese-American Reading Room at Guangxi University, in Guangxi SAR, PRC. Dr. Masterson has a Ph.D. in Political Science from the University of Cincinnati. His research focuses on international relations theory and research methodology with specialization on international political economy. His research interests include the international relations of East Asia in general and analyzing the nexus between Chinese foreign and economic relations with the world in particular. His publications in China Information and Commonwealth Journal of Political Science examine the effects economic factors have on Chinese political relations with its neighbors from 1987-2001, and the world from 1990-2005 respectively. Dr. Masterson also co-published an article titled "Thirty Years of Growth in Kentucky's Prison Population and Spending: A Multi-causal Analysis" in the Contemporary Journal of Anthropology and Sociology. This research seeks to advance our understanding of the factors that contribute to the state's prison population growth and capital expenditures for state corrections. Dr. Masterson's latest published work is a book chapter in Revolutionizing the Interaction between State and Citizen through Digital Communication that examines the double-edged sword that growth in social media in China poses to the ruling elite. He currently is coauthoring a manuscript seeking to advance our understanding of the influence that bilateral meetings with China's senior leaders have on its foreign relations. Professor Masterson teaches courses on

International Relations Theory, Current World Problems, US Foreign Policy, Asian Politics, Research Methods and Comparative Politics.

**Nicholas Ravotti** is an attorney licensed in various tribal courts int he United States. His practice focuses primarily on litigation, natural resource law, education, and housing. He represents tribal governments and tribal entities in state and tribal courts.

**Colin Rule** is Vice President of Online Dispute Resolution at Tyler Technologies; former Director of Online Dispute Resolution, eBay/PayPal.

**Mylla Sampaio** is a Bachelor of Laws (LLB) by the Universidade Federal do Maranhão (2018). Master of Laws (LLM) by the Universidade Federal do Maranhão (2020). Research Fellow at the University of Salamanca. Research Fellow at Núcleo de Pesquisa Cultura, Direito e Sociedadeate the Universidade Federal do Maranhão.

**Mark Wilson** is Senior Product Manager, Online Dispute Resolution, Tyler Technologies.

# Index


#Yosoy132 49-50, 54-58, 61-65

## A

appeals 80-81, 94-95, 105-111, 113, 116, 118, 143
Application Programming Interface (API) 122
Arab Spring 41, 149, 164, 181

## B

Blogosphere 142, 164, 181

## C

censorship 33, 52, 68, 140, 147-150, 168
Checks and Balances 2, 7, 17-18, 24
civil society 5, 10, 17-18, 30, 42, 44, 46, 52, 56-58, 146, 186, 191
Classical Democracy 46
Cloud 101-104, 109, 114, 116, 120, 122
Common Good 4-5, 7-8, 13, 18, 24-25, 44
Constitutional amendment 194-195
Cyberactivism 146, 181

## D

Democracy 2-13, 15, 17, 20, 25-27, 34, 39, 41-43, 46-47, 50, 52, 54, 64-65, 69-71, 79, 82, 85, 92, 167, 183, 185
Democratization 5, 8-10, 12-13, 56, 58, 62
Digital or E-Democracy 46
Direct Democracy 46-47
dispute 60-61, 79-81, 94, 109, 112-114, 118, 123, 150-151, 156-158, 165-167, 192

## E

E-Government 40, 97, 123
Electoral reform 182-184, 188, 192, 194, 200
electronic voting 182-184, 186, 196, 198-200

## F

Facebook "Like" 81, 92
First Amendment 68-71, 75-76, 79, 81-83, 133
freedom of assembly 69, 71
freedom of speech 26, 44, 69-73, 76, 78-79, 82, 84-85, 125

## I

Indian Tribes 125-129, 131-132, 134

## J

Jasmine Revolution 164, 167, 181
John Stuart Mill 26, 32, 36, 43

## L

Law 2, 7, 9, 17-18, 25, 27, 36, 41, 69, 77-78, 81-82, 84-87, 93, 106, 108, 113, 124-129, 131-137, 155, 163, 166, 182-183, 185-188, 190-196, 199-200
Legal Research 124-126, 133-136
Legitimacy 1-3, 5, 10-11, 14-20, 25-26, 30-31, 33, 39, 43-44, 127, 140, 142, 181
Liberty (Mill's) 47

## M

Mainframe Technology 97, 123
Marketplace of Ideas 70, 75, 82-83, 86, 92
Microblog 143, 158, 163, 166, 181
mixed system 182, 184, 188-189, 195-196
mobilization 49-50, 52-53, 61, 63-65, 142, 146, 168

## N

negotiation 117-118

## O

ODR 109, 114
Online Dispute Resolution 94, 109, 114, 123

## P

participation 5-7, 9, 26, 29-30, 36-38, 43-44, 49-51, 53-54, 64
PRC 150, 157
Prior Restraint 73-75, 93
Protest Leader 144, 181

## R

Representation 2, 7, 12, 15, 17-18, 25, 43-44, 64, 182, 189, 193, 199
Representative Democracy 47, 183
resolution 29, 94, 109, 111, 114-116, 123, 195
Rule of Law 7, 9, 17-18, 25, 36, 185

## S

Separation of Powers 17-18, 25
Sichuan Earthquake 140, 150, 152-153, 155, 161
Social Media 2-3, 5-6, 11, 19, 26, 31-33, 37-44, 49-54, 57-58, 60-65, 68, 70-71, 79-80, 82-86, 92, 139-152, 154-156, 159, 164-166, 168-169, 181
social movements 57-58, 62-63, 65
social participation 49, 64
Software as a Service (SaaS) 123
Speech 26, 44, 52-53, 58, 65, 68-76, 78-87, 92-93, 125, 142-143, 167-168
Sunk Cost 123

## T

tax 40, 94-95, 105-110, 113, 118, 131
Time, Place, and Manner Restriction 73, 93
transparency 10-11, 15-16, 26, 29, 33, 36-39, 43-44, 56, 72, 85, 111, 116, 163, 186, 196, 198
Tyranny (Modern) 47

## U

Uptime 104, 123

## W

Will of the Nation 35, 47
wireless interruption 72, 74

www.ingramcontent.com/pod-product-compliance
Lightning Source LLC
LaVergne TN
LVHW081324300926
849161LV00008B/1868